A Sun is NOT a Gas Sphere

by Rolf A. F. Witzsche

Contents

3

About the Illustrated Science series
On the Ice Age and Climate Change
and the book
A Sun is NOT a Gas Sphere
Book 1 of the series: 'Cold' Plasma Fusion Powers the Sun

It is not possible for the Sun to be a sphere of hydrogen gas that is powered by nuclear fusion from within. Many epicycles have been invented to make the concept seem plausible. All evidence points to the only possible fact that the Sun is a plasma star that is externally powered with plasma fusion occurring at its surface. The difference between the two models, which are opposite in nature, is enormous.

Ironically, the Hydrogen-Sun model, which is not physically possible, is being taught in the schools, while the Plasma Sun model, for which evidence exists, is being hidden from view. The reason may be that it is difficult for people not familiar with plasma physics, to see the universe as a Plasma Universe, even while research reveals that 99.9999% of the mass of the universe exists in the plasma state; in the form of free-flowing protons and electrons. In plasma fusion on the surface of a sun, plasma is combined into atomic elements, with energy being generated in the process. The atoms for all the planets and gases in a solar system are synthesized at the surface of the Sun, which actually powers the solar process.

The synthesizing plasma fusion is presently enhanced for our Sun by electromagnetic 'Primer Fields' focusing interstellar plasma onto the Sun in a highly condensed manner. When the plasma-focusing system becomes inactive, below the required threshold conditions, the Sun reverts to a type of cosmic default level with 70% less energy being radiated, and higher rates of solar cosmic-ray flux are being experienced.

At the present rate of plasma diminishment being experienced, the solar activity phase-shift threshold to the next Ice Age period may be crossed in 30 years, or in the 2050s, most likely. With the primer-fields system gone inactive by then, the climate on Earth will get 40 times colder than the

Little Ice Age in the 1600s had been. Ice core evidence promises that. Without the needed preparations for human living in such an environment, 99% of humanity would die of starvation, both by the cold, and by CO2 depletion that diminishes agriculture, as more CO2 becomes dissolved into the sea.

With the 'Primer Fields' being critical for our very existence, the exploration of them is likewise critical.

In the Little Ice Age, between 10% and up to 30% of the populations in Europe had perished by starvation. The last Big Ice Age was evidently vastly harsher. Only 1-10 million people emerged from it alive. That's all we had after 2 million years of development. We want to do far better this time around; and we can, with large-scale technological infrastructures for our food supply. But will we create them? Will we get the job done in the 30 years that we still have left before the Ice Age starts anew? Will we even consider it? And how certain are we that the phase shift to the next glaciation period will begin, as the evidence suggests, in the 2050s? We have no slack on this front. Should we fail us on this absolute front, we would be committing suicide.

Numerous fields of evidence tell us that the next Ice Age is near. That's where the truth begins. Most of the evidence was discovered in the 1990s and thereafter. Some evidence is measured in ice cores; some is measured in space, by satellites. Some measurements are also made on the ground in terms of measurements of the Earth's magnetic-pole drift observed in northern Canada. All of this is seen combined with high-energy physics experiments at a leading national laboratory, and is also explored in the small in static experiments.

So, what will the answer be? Will we move with the evidence? Or will we lay ourselves down to die by default?

It takes an independent researcher to brake the taboos that have kept mainstream cosmology imprisoned, increasingly, during the past century, even while what is regarded as taboo is known to be wrong.

The Illustrated Science series is intended to open the scene beyond the threshold of accepted taboos, to where the actual physical evidence speaks for itself.

The scope of the existential challenge that the Ice Age brings with it, takes astrophysics out of the academic domain and places it into the foreground as one of the most-critical issues of our time. The big Climate Change events that have already worldwide effects are mere fringe effects in the flow of the ever-changing cosmic dynamics. The big effect, when the Ice Age begins anew, promises to be caused by a dimmer and colder Sun. The loss of 70% of the Sun's radiated energy defines our climate future that begins in the near term.

Sure, we can live with all that by creating new platforms for agriculture that are able to operate under Ice Age conditions. But will we do it? The task is enormous. Or will we fail ourselves on this front? We have no reason to allow us to fail. We have the materials and energy resources on hand to accomplish everything that is required for us to continue to live in an Ice Age World. But will we do it? The big question that never goes away, therefore, is; will we develop our inner resources as human beings sufficiently to get the job done, and to get it done in time? Or will we do nothing, ignore the challenge, and condemn our children and one-another to an agonizing death by starvation? That's the choice.

Towards meeting the inner challenge, I have created the epic series of novels, The Lodging for the Rose. And further, towards meeting the science challenge, I have produced numerous research books and several dozen exploration videos that the Illustrated Science series is modeled after. The work is the result of a quarter century of research, for which numerous elements of evidence in related fields came to light during the timeframe of my research.

It is my hope that the work that went into all of these projects will help in some degree - for humanity that we are all a part of - to write itself a ticket to have a future.

High-resolution color images, of the images in this book, can be obtained at www.iceagetheatre.ca

Let me surprise you
A confusion of theories abounds, about what our Sun is, and how it operates.
Primarily the confusion is divided into two groups of theories based on completely opposite concepts.

The Big Bang Creation ideology

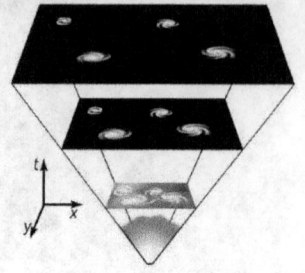

The Big Bang creation myth refuted by the electric solar fusion model

One concept is centered on the Big Bang Creation ideology. It speaks of a gigantic explosion 13.8 billion years ago, in which all the matter and energy of the entire universe was created at a single place in the first three minutes. The explosion is deemed to have furnished all the dust of the cosmos that was driven apart by the shockwave into all directions, where it condensed by gravity into planets, stars, and galaxies.

The electric cosmology

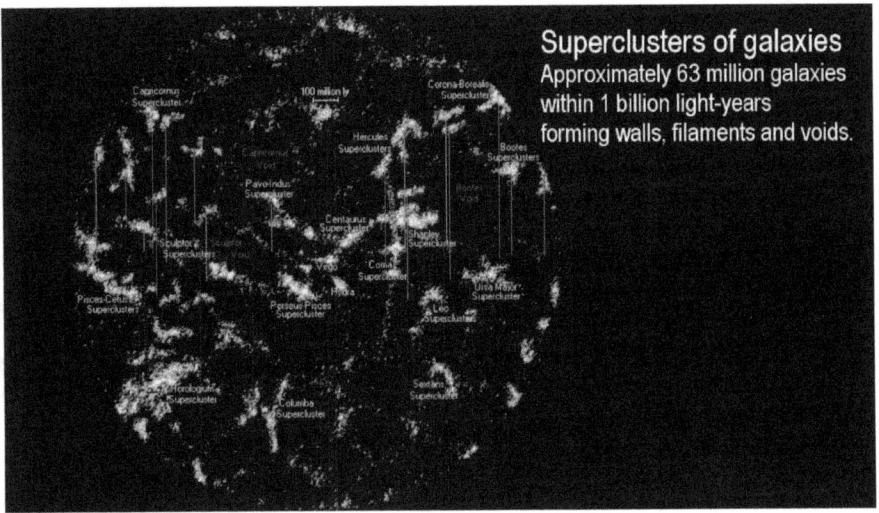

Superclusters of galaxies
Approximately 63 million galaxies within 1 billion light-years forming walls, filaments and voids.

The opposite concept is the electric cosmology that speaks of a boundless universe that is pervaded by endless streams of plasma that interconnects all the galaxies and powers all the stars within them, including our sun. In the electric cosmology, plasma, which caries electric potentials, is deemed the lifeblood of the universe. The two opposite concepts have each forged opposing trends of theories that affect how we respond to the dynamics of our Sun and its operating principles. They also affect how we develop our future.

It is here, in this context of our theories affecting the future, where things begin to get serious.

The effects of cosmological theories have an enormous impact on the future of humanity, by the way we respond to the effects. The effect that we face in the near future promises to be so immense that it takes the duality of the scientific perceptions out of the realm of merely academic significance, and gives it an existential significance.

The long suspected resumption of the Ice Age

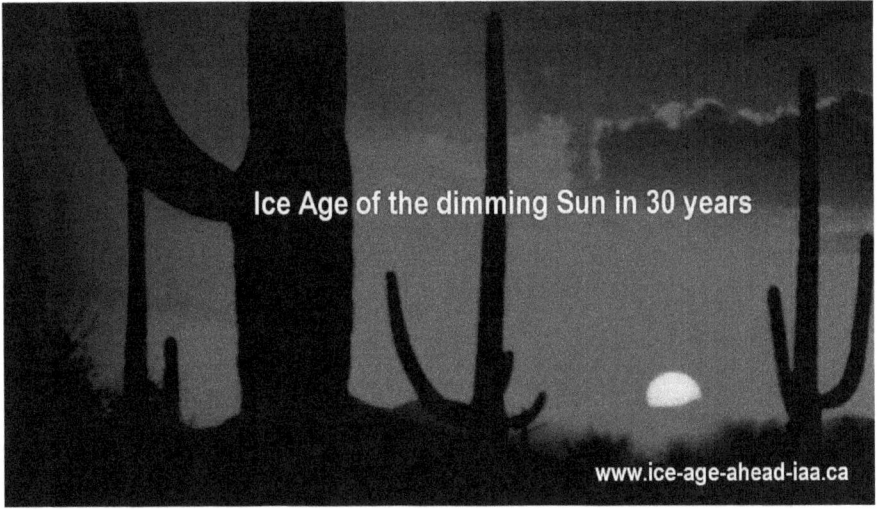

Ice Age of the dimming Sun in 30 years

www.ice-age-ahead-iaa.ca

For example, the long-term exploration of the principles of plasma dynamics, applied to cosmic dynamics and solar dynamics, has yielded the astonishing recognition, with a high degree of certainty, that the long suspected resumption of the Ice Age that is deemed to be still thousands of years in the future, is much nearer and will likely start in the 2050s with the Sun going inactive. It will unfold extremely rapidly under a 70% dimmer and cooler Sun.

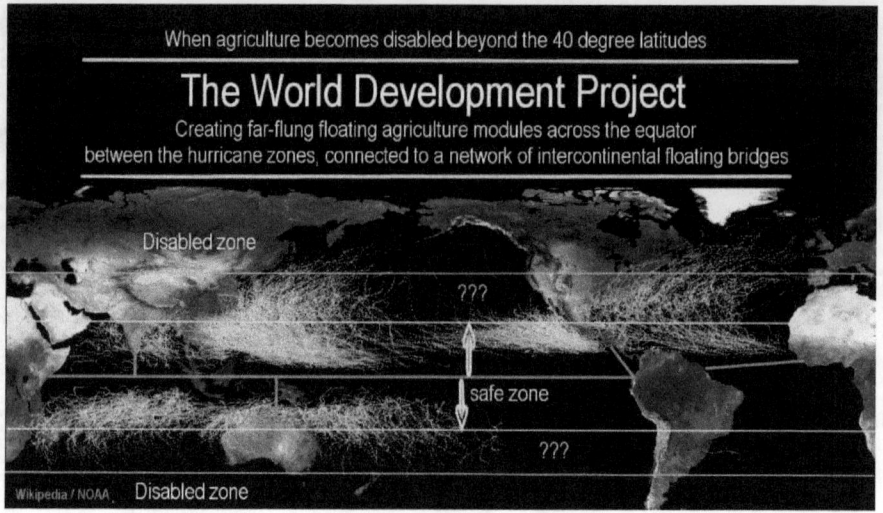

With the potential transformation of our world on this gigantic scale looming on the near horizon, the determination of what is truth is no longer a mere academic concern, but is a concern that will determine the future existence of humanity. The transformation of our world of the magnitude that arises from a 70% dimmer Sun, will render almost all countries above the 40-degree latitude unsuitable for agriculture if not completely uninhabitable. This means that much of the world's present agriculture will have to be relocated into indoor facilities or be placed afloat across the tropics, as little suitable land exists in the equatorial regions.

Gigantic infrastructure development

It is self-evident that the type of gigantic infrastructure development, as will be required - complete with floating cities to service the floating agriculture - won't be contemplated, much less be built, unless the science division is healed.

Obviously, of two opposite science concepts, only one can be real. But why do we have two opposite concepts? Isn't science a quest for understanding what is real? This is true, but only to the point where politics enter the scene.

Ptolemy
being guided by philosphy

Margarita Philosophica by Gregor Reisch, published in 1508

Throughout the pages of history, science has been 'guided' with the payola and other means to serve the doctrines of the various empires, which typically have the power to have their objectives met. On this train, truth falls by the wayside. It always has. This applies to cosmology too.

The train of politically 'guided' science

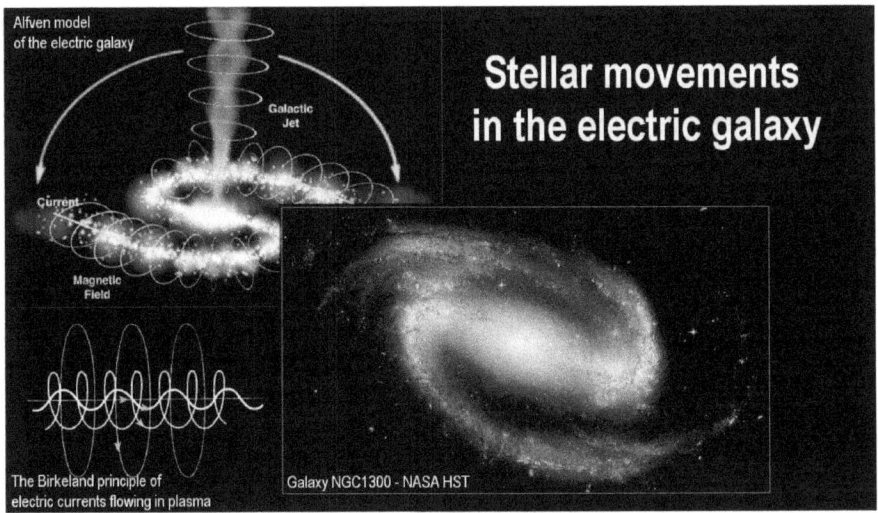

Alfven model of the electric galaxy

Galactic Jet

Stellar movements in the electric galaxy

Current

Magnetic Field

The Birkeland principle of electric currents flowing in plasma

Galaxy NGC1300 - NASA HST

While the train of politically 'guided' science goes far back into history and has many tales attached, the counter science in cosmology appears to have been invented to counter the breakthrough discoveries in plasma physics by the Swedish, 1920 Nobel Price winner in Physics, Hannes Alfven.

The timing suggests that the Big Bang Creation theory was developed, and was massively promoted, as a counter-theory against the plasma universe.

The plasma universe offers an unlimited electric-energy future to humanity. The recognition of it would scrap the value of the private ownership of the world's energy resources that is one of the pillars of empire. In the Big Bang cosmology protects that pillar. Under the Big Bang doctrine, plasma is deemed not to exist. Of course it does exist. A vast body of physical evidence testifies that it does exist.

Artist imagined
black hole

NASA/JPL

Inversely, no real evidence exists for the exotic Big Bang concepts that are hugely played up and promoted as real, such as black holes, dark matter, and stellar explosions.

Deep in the Big Bang cosmology

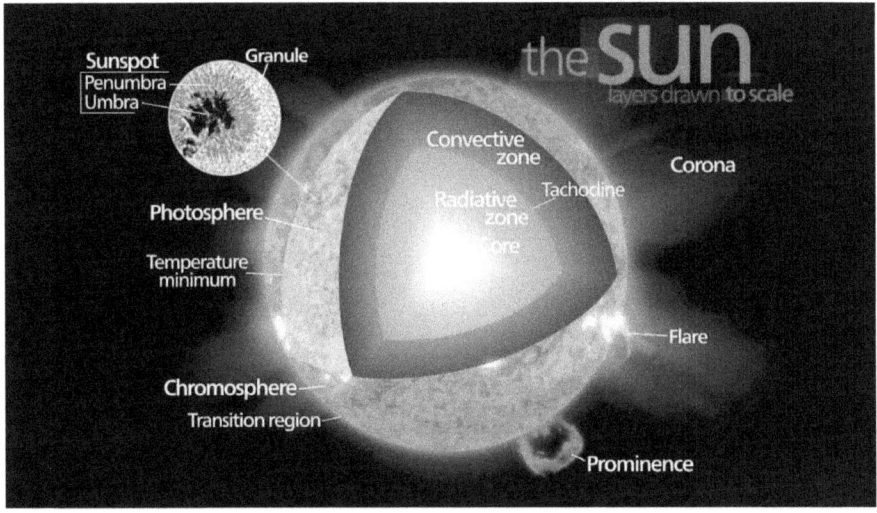

Deep in the Big Bang cosmology, where only gravity is allowed to be considered as a causative force, we find also the theory promoted that our Sun is a gas sphere, powered by nuclear reactions occurring in its core, which are deemed to fuse hydrogen atoms into helium atoms.

The theory of the internally powered Sun, renders our Sun a universal constant for all climate considerations, which literally closes the door to rational ice age concepts.

The trap that has been created with this prevents humanity from preparing its world for the next Ice Age to come, that may begin in 30 years. On this train of the Big Bang theory the depopulation objective is well served, which is a core policy within the oligarchic system of empire. The stated objective of the masters is, to reduce the population of humanity to less than 1 billion people, as required for the stability of a feudal system.

This is what stands behind the science duality. But how do we get out of this trap? Were do we go from here? The answer must be that we dare to explore what is real.

The Earth stands as a silent testimony

NASA - Earth from Apollo 16
wikipedia

Ironically, it is the Earth itself that weighs heavily against the Big
Bang theory. The Earth stands as a silent testimony that its atoms
were created brand new when the Earth itself was formed. This
scientifically proven fact, backed with hard evidence from a wide
range of sources, turns the Big Bang theory, and everything that is
built on it, into a very imaginary fairy tale.

Since the fairy tale has spawned numerous theories and inspired
many opinions, it becomes necessary, therefore, in exploring the
truth, to separate what is demonstrably real, from the landscape of
educated opinions and cultivated illusions, and that one does this in
a comprehensive manner. The reason is that ultimately, the subject
of truth, is a single package.

Outside the topics of fairy tales of the Big Bang

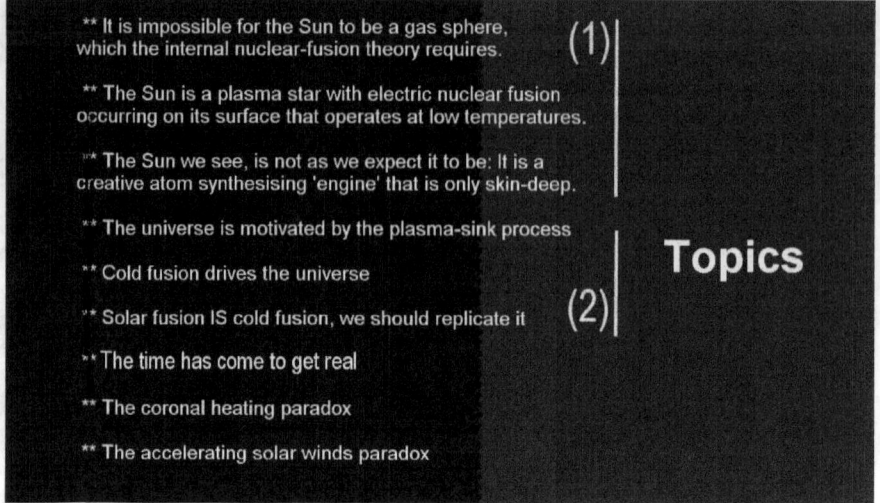

In the course of the exploration the video introduces a number of revolutionary concepts that may seem surprising, but which are critical for a rational understanding of the dynamics of the solar system outside the topics of fairy tales of the Big Bang cosmology. This means that the resulting presentation of the nature of our Sun as an energy source, won't be of a type that is taught in schools, institutions, and is presented in science documentaries for the television audiences.

Since the field of exploration that is presented here, has become largely unknown, but covers a number of related concepts, the video exploration is being presented as a series of 7 parts. The evidence with which the Earth refutes the Big Bang cosmology is presented in Part 2, which thereby becomes a part of the evidence for the plasma Sun, that has electric nuclear fusion occurring on its surface where its energy-radiation originates. But before we can get to this, Part 1 of the series is needed to establish what the Sun really is.

The internal-nuclear-fusion theory of the Sun, which is the generally accepted theory, has many flaws built into it, while no evidence actually exists that exclusively supports the theory. None whatsoever! That's shocking, isn't it? All historic and visible evidence supports instead the recognition of the Sun as a plasma star that is externally powered with cold-fusion nuclear synthesis occurring at its surface. This affects our climate, economics, politics, and how we relate to one-another as human beings.

When the truth becomes known, the world is changing. The old theories no longer apply. Recognized evidence discredits them.

Disturbing to those who cherish the illusion

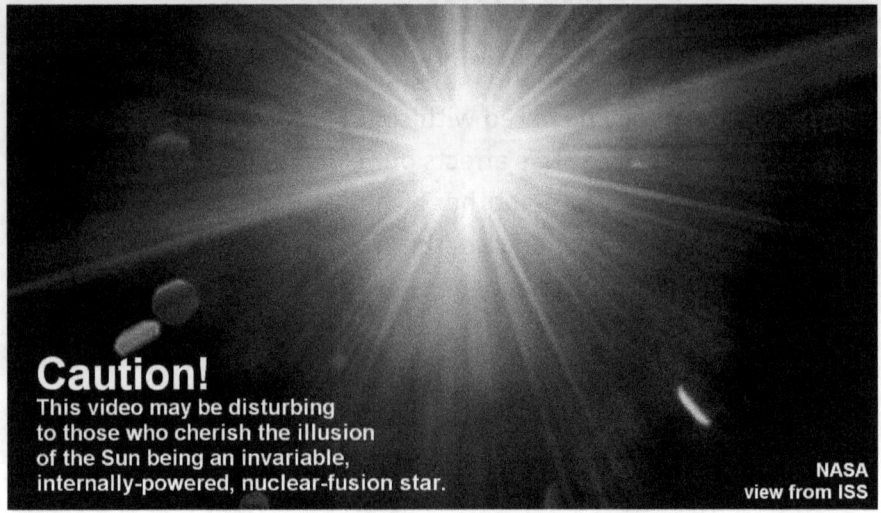

Caution! This video may be disturbing to those who cherish the
illusion, of the Sun being an invariable, internally-powered, nuclear-
fusion star.

Topics of truly gigantic plasma structures

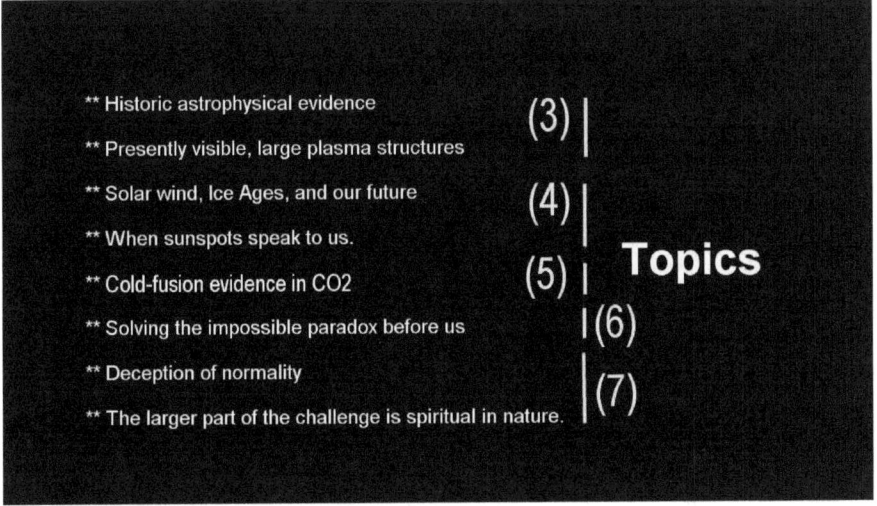

Part 3 presents evidence in historic events of large plasma structures in the sky, that were evidently visible in the past, but are no longer visible in our electrically weak times. Inversely, it also presents topics of truly gigantic plasma structures that were never visible, until recent times, when modern instrumentation made them visible to us. All of these prove that we live in the face of an electrically powered Sun.

Part 4 deals with the connection between the solar wind and ice ages in the electric universe.

Part 5 deals with the largest historic electric events in the solar system, their impact on life on Earth, and the imperative for us to save the future.

Part 6 deals with evidence for the near impending end of the current interglacial period, and the start of the next Ice Age in 30 years, and the certainty of the transition.

Part 7 deals with the biggest question that the entire series is leading up to. The question is whether we will respond to the scientific imperatives imposed on us by the future, to protect our existence.

26

Will we build the 6,000 new cities that we need?

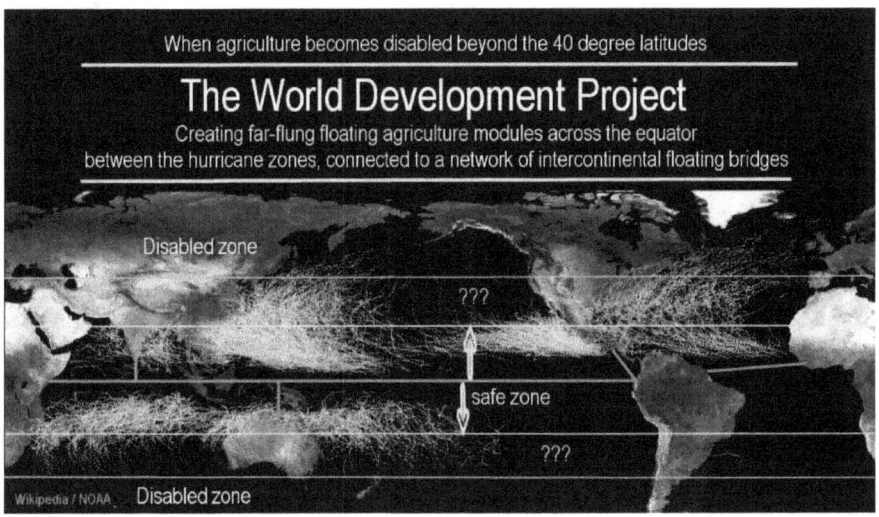

We have the materials, technologies, and energy resources on hand to build the worldwide infrastructures with which to secure the human landscape in the harsh time ahead when the Ice Age starts anew and large parts of the world become uninhabitable.
But will we use the resources we have? Will we built the 6,000 new cities that we need to enable the relocation of most of the great nations on the Earth? Will we do this and live? Will we built the 20,000 kilometers of intercontinental bridges along the equator, and the millions of acres of floating agriculture, that the bridges would connect to?

Will we create a new world for us?

Ice Age of the dimming Sun in 30 years

www.ice-age-ahead-iaa.ca

Will we do all of this and create a new world for us for a richer living under harsher conditions? Or will we do nothing and allow ourselves to be blown away with the winds of the cycles of the universe?
Part 7 deals with the difficulty in answering these questions.

The largest part of this spiritual challenge

In this context the entire scene becomes no longer merely a physical challenge, or a technological question, or even a science issue, but becomes a spiritual challenge. The largest part of this spiritual challenge will ultimately be, whether we will rouse ourselves to regard one another as human beings, with enough love for one another as the brightest diamond in the landscape of life, that the currently faint spark of love for one another becomes a fire of universal love.

In the wonders of our humanity

The Greatest Miracle on Earth
a human being
born in the image of God
reaching for spiritual light
to understand the universe
and itself

Then, when we get to this point, the path will be free for the brightest future imaginable. And for this imperative too, we have the resources already at hand in the wonders of our humanity.

Our Electric Cold Fusion Sun (Part 1) A Plasma Star

The Sun is not a gas sphere

The Sun is not a gas sphere. It cannot be that.
It is impossible for the Sun to be a gas sphere, as the theory of the internally powered Sun requires it to be. But why is this impossible? The answer is simple. It couldn't operate in any other manner, because its operational principle is simply the most efficient one there is. Nor is it self-powered. It is powered by a principle that involves almost the entire universe. Physically, it is powered by plasma. Plasma is the life-blood of the universe. It is electrically charged. This makes it powerful. The electric force is 39 orders of magnitude stronger than gravity. The inclusion of plasma opens the empty box of conventional astrophysics where only gravity is deemed to rule and 99.999% of the universe is deemed not to exist.

Image: Plasma streams

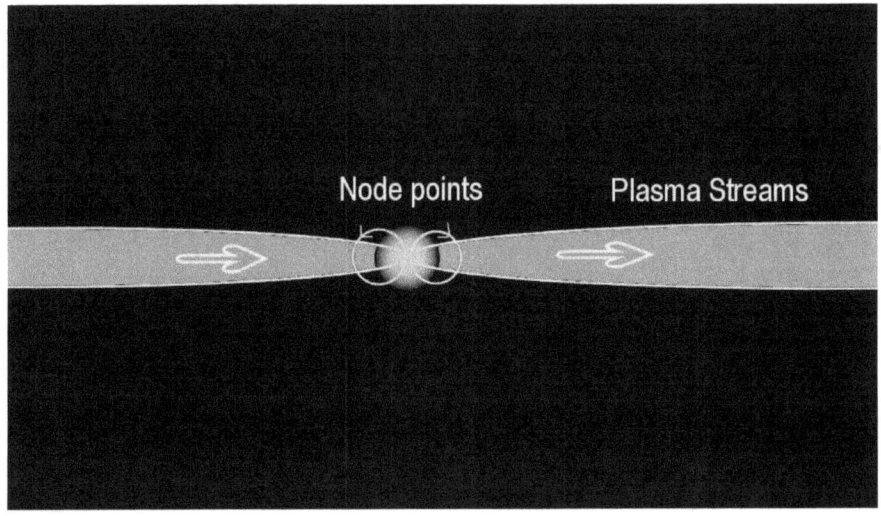

Image: Plasma streams

Image: galaxies as node points

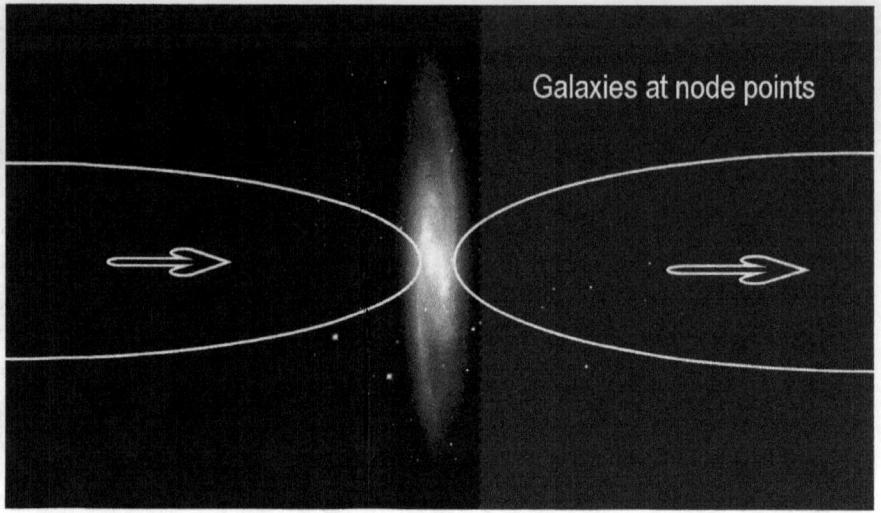

Galaxies at node points

Image: galaxies as node points

Image: The Milky Way, a node

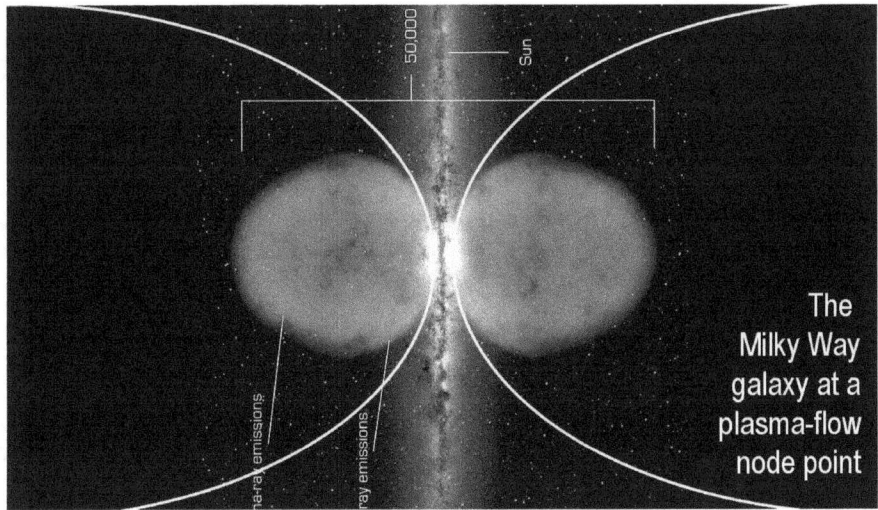

Image: The Milky Way, a node

Image: The Milky Way is plasma powered

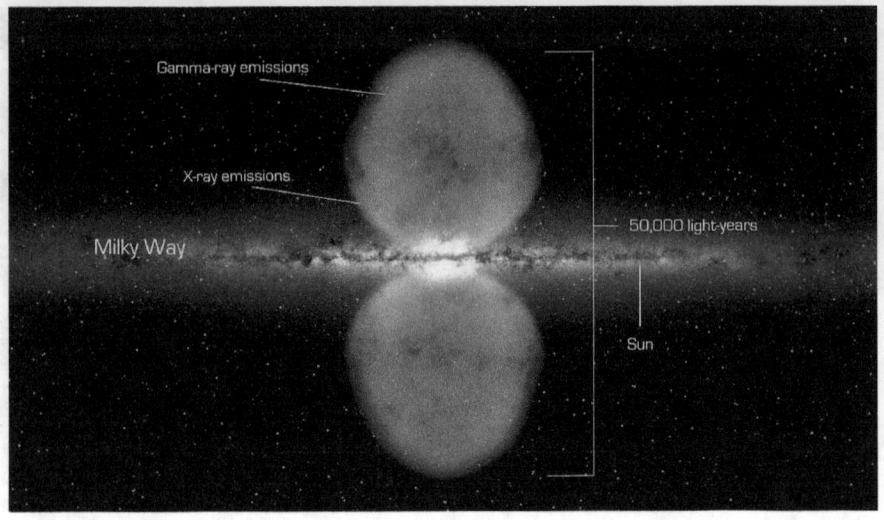

Image: The Milky Way is plasma powered

Image: Milky Way plasma domes

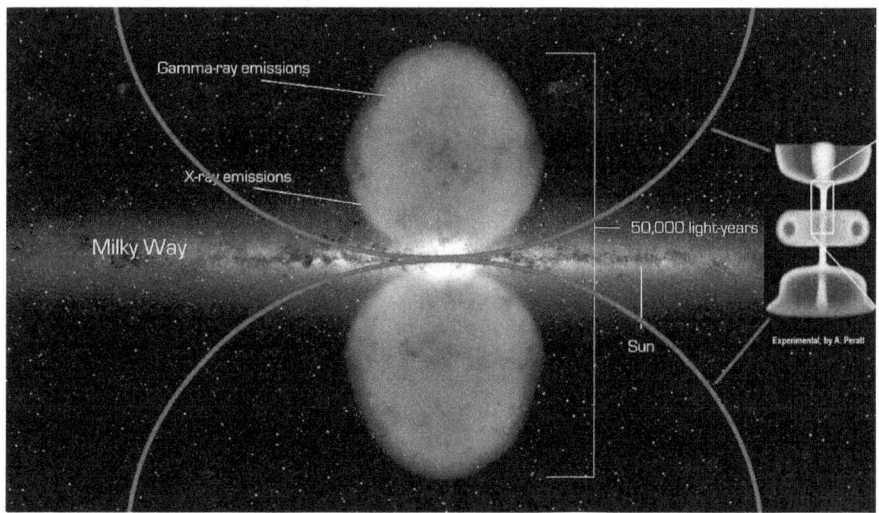

Image: Milky Way plasma domes

Image: The Milky Way in comparison

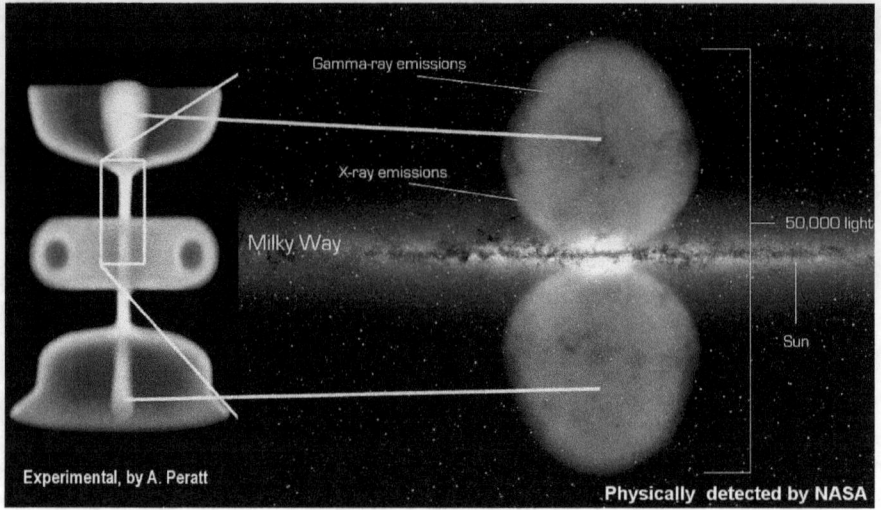

Image: The Milky Way in comparison

Image: Stars at node points

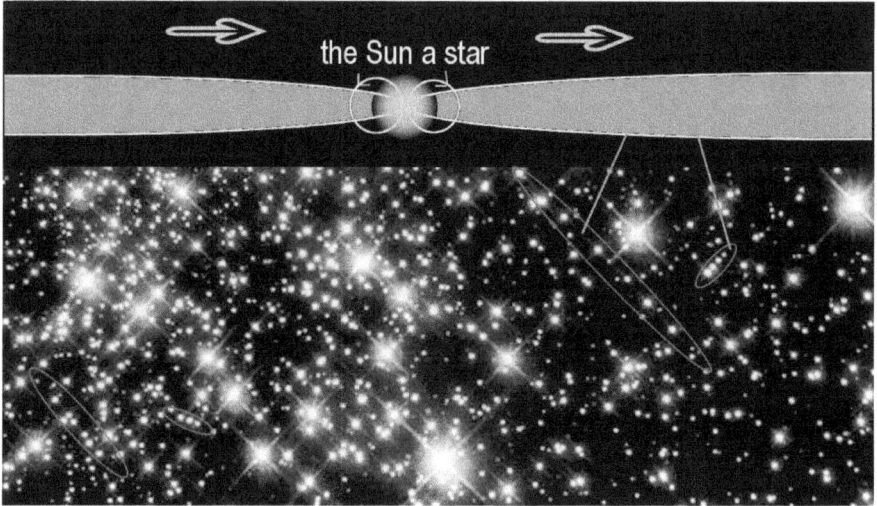

the Sun a star

Image: Stars at node points

Image: Plasma focused onto the Sun

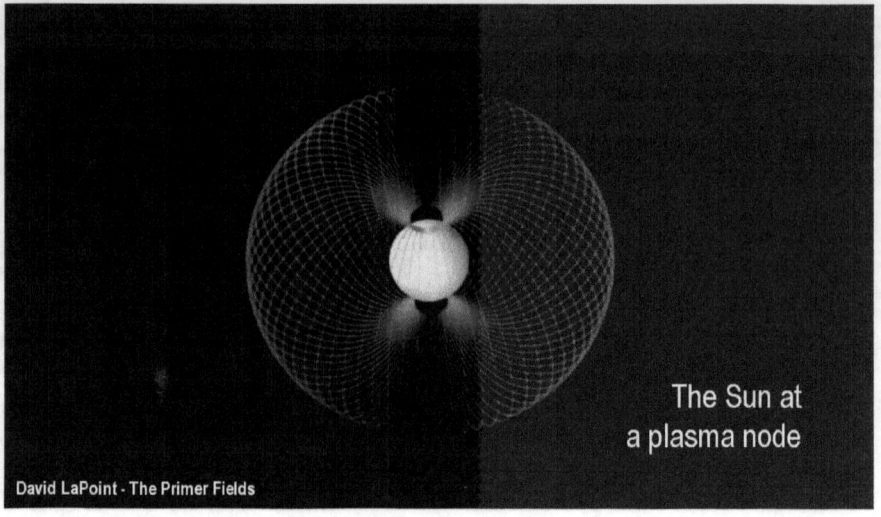

The Sun at
a plasma node

David LaPoint - The Primer Fields

Image: Plasma focused onto the Sun

Image: Powered by interstellar plasma streams

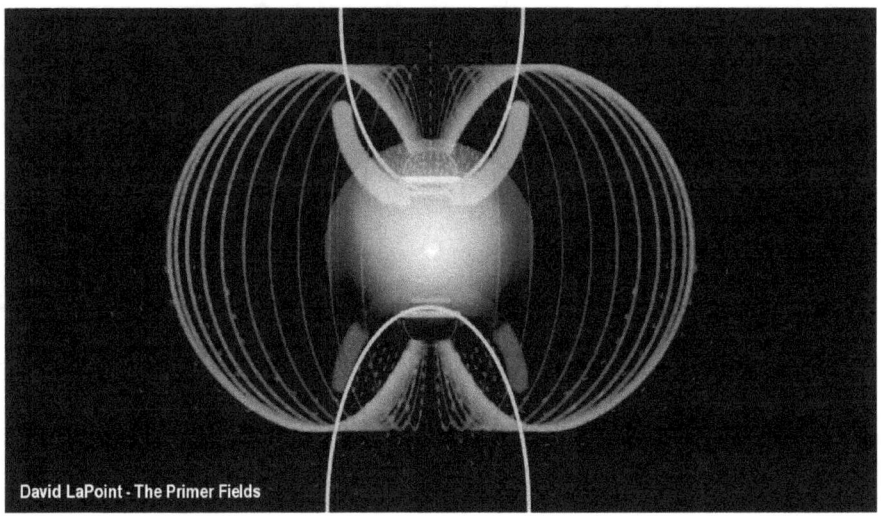

Image: Powered by interstellar plasma streams

Image: LaPoint and Peratt discoveries

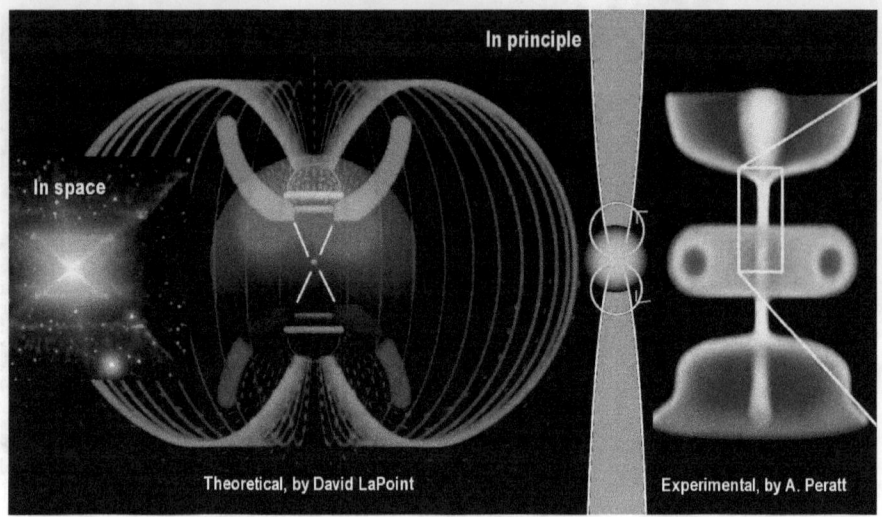

Image: LaPoint and Peratt discoveries

Image: LaPoint, artificial Sun

Image: LaPoint, artificial Sun

Image: The Milky Way lookalike

Image: The Milky Way lookalike

Image: The Sun located in the Milky Way

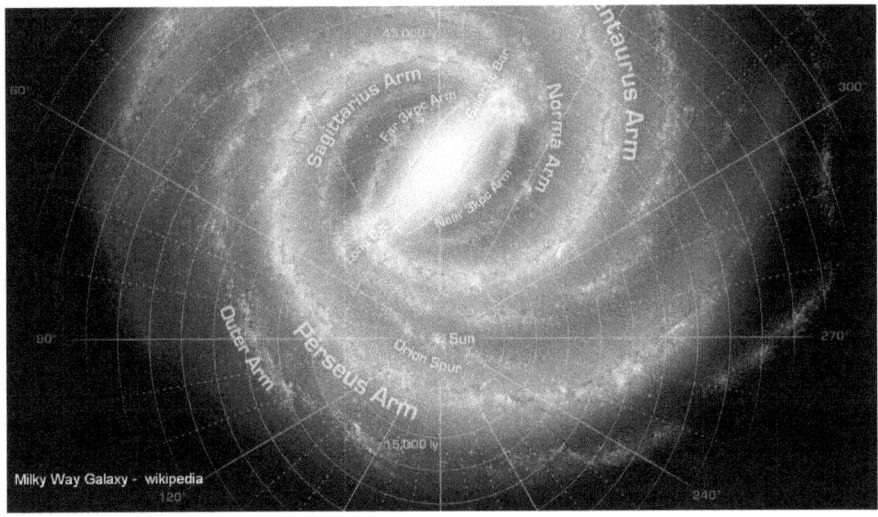

Image: The Sun located in the Milky Way

Image: The x-ray Sun

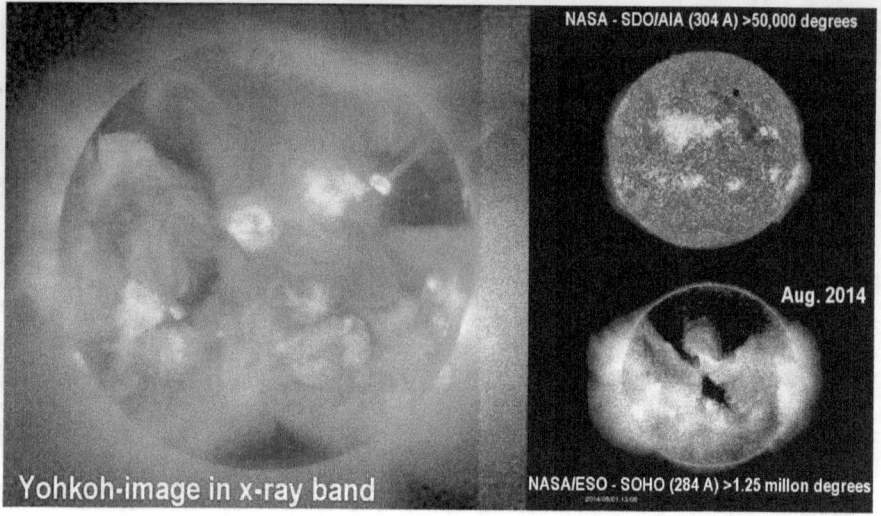

Image: The x-ray Sun

Image: Star-size comparison

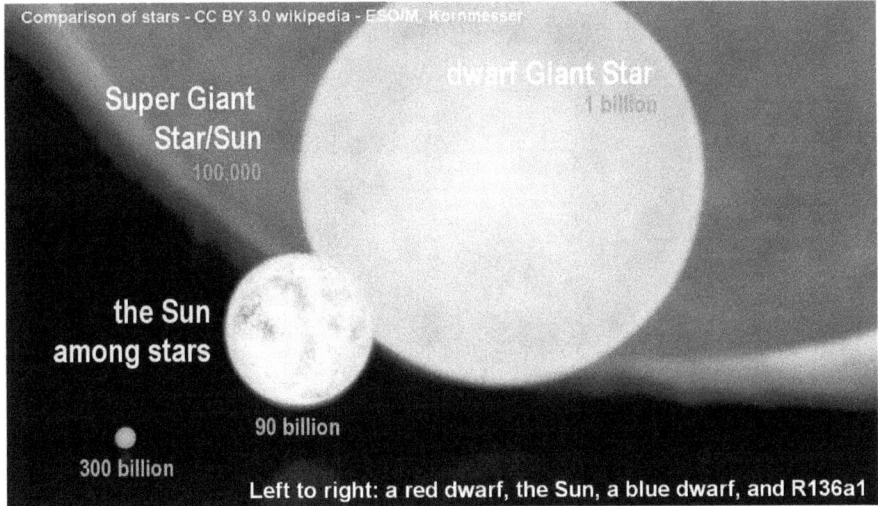

Image: Star-size comparison

47

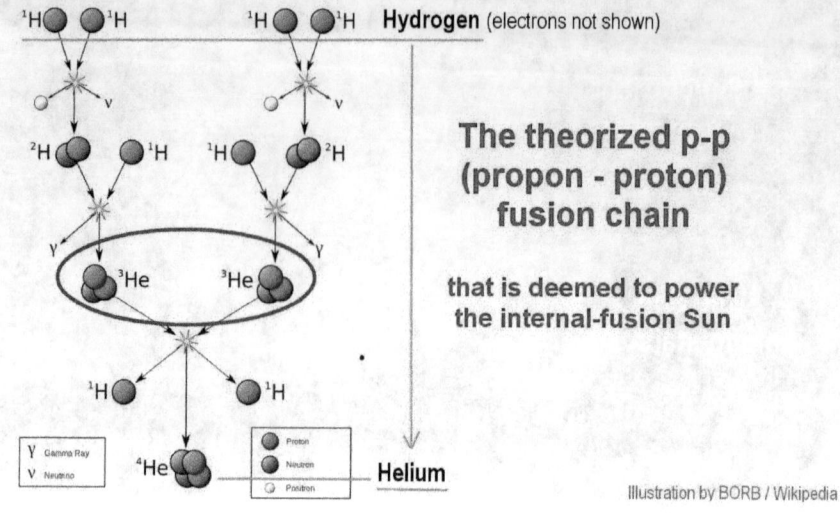

The theorized p-p (propon - proton) fusion chain

that is deemed to power the internal-fusion Sun

Illustration by BORB / Wikipedia

In real terms, the conditions for nuclear fusion to occur deep inside a sun, that force hydrogen atoms to fuse into helium atoms, which the theory is based on, do not exist at all. A sphere of hydrogen gas of the size of the Sun cannot exist. Its weight would crush all atoms in its core, much less enable the building of bigger ones. In addition, the theorized helium-3 fusion-stage is the hardest fusion to achieve, because of the much larger Coulomb Barrier of helium-3 that needs to be overcome for such proton-heavy atoms to fuse. In a NASA related helium-3 fusion experiment, it took a million times greater energy input to force the fusion than the energy that the fusion had generated. A sun cannot operate on this basis.

In real terms, there is no need for an internal nuclear-fusion power process to happen, for a sun to radiate light and heat. The real solar nuclear fusion process is much simpler; more certain; and more powerful; and in addition the fusion occurs right on the surface of the Sun, where it counts.

The solar-wind particles are plasma

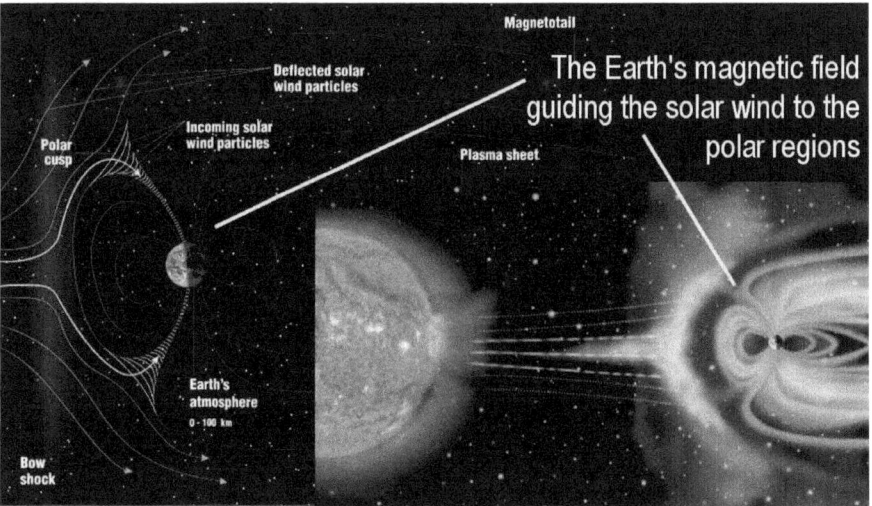

Deflected solar wind particles

Magnetotail

The Earth's magnetic field guiding the solar wind to the polar regions

Incoming solar wind particles

Polar cusp

Plasma sheet

Earth's atmosphere
0 - 100 km

Bow shock

The solar-wind particles are plasma. They carry an electric charge by which they interact electrically with the Earth's magnetic fields. By this interaction, they are guided magnetically to the polar regions where they encounter the atmosphere and create their highly visible light show.

Guides interstellar plasma onto our Sun

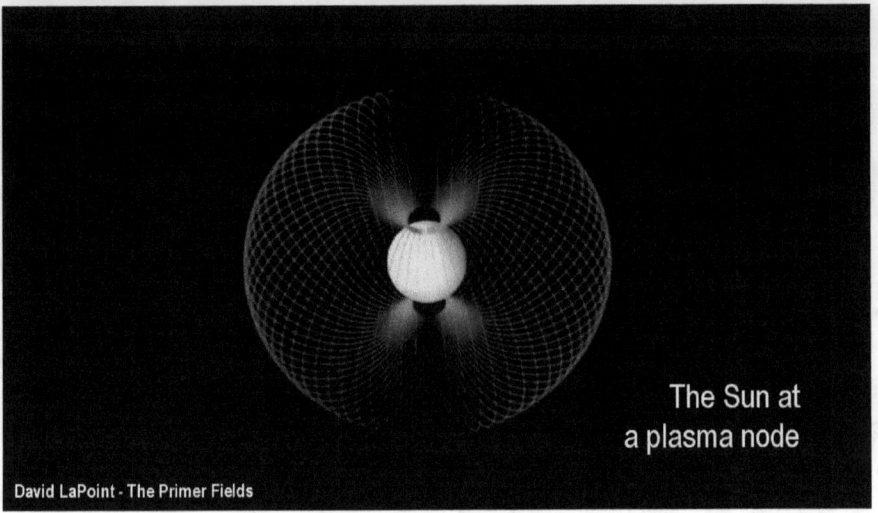

The Sun at
a plasma node

David LaPoint - The Primer Fields

The same type of process, only on a much larger scale, also guides interstellar plasma onto our Sun itself.

In the case of the Sun

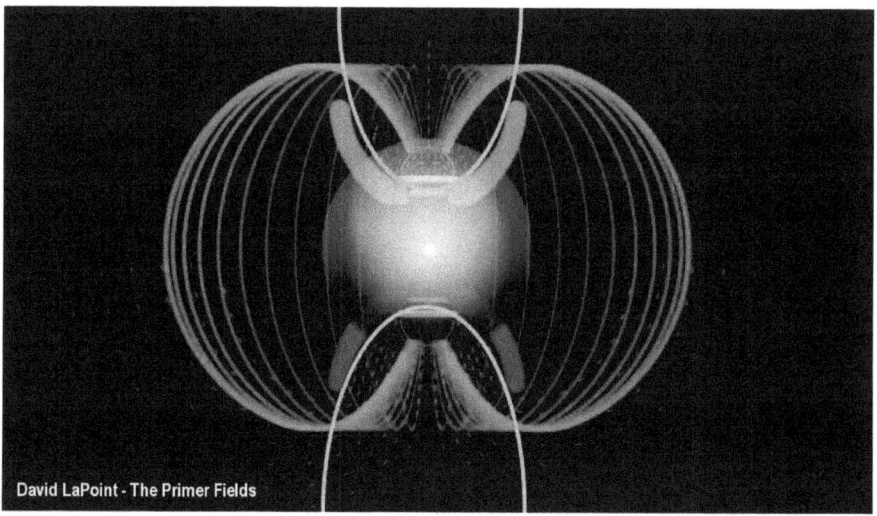

David LaPoint - The Primer Fields

The difference is, that in the case of the Sun, the plasma comes from a far more-distant source. It is drawn to the Sun from interstellar space, in the form of long-distance plasma streams.

Plasma is electricity

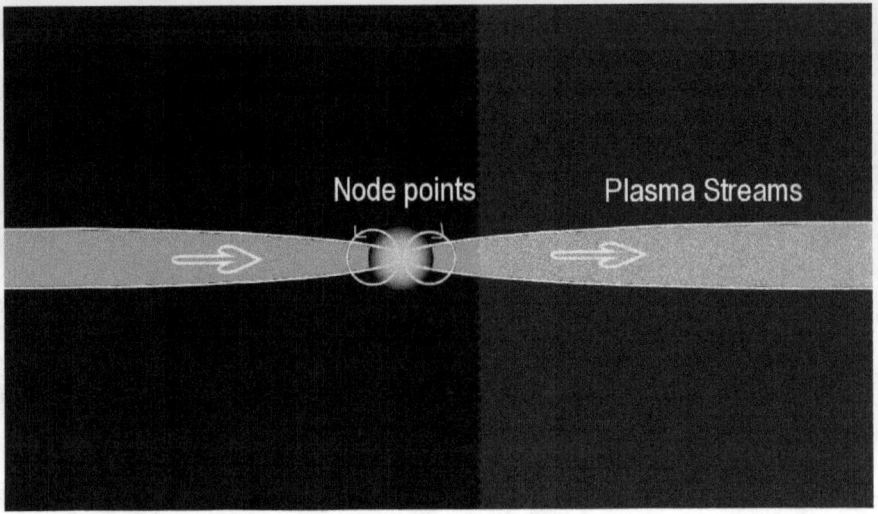

Since plasma is electricity, the flowing plasma currents create their own magnetic fields around them that pinch the flowing plasma into ever tighter confinement. The resulting concentrated magnetic fields, in turn, amplify the magnetic field of the Sun with their own magnetic structure.

The Sun becomes surrounded

The Sun at
a plasma node

David LaPoint - The Primer Fields

The result is that the Sun becomes surrounded with a densely concentrated sphere of plasma. The plasma becomes so dense that the resulting 'plasma light-show' extends right around the Sun.

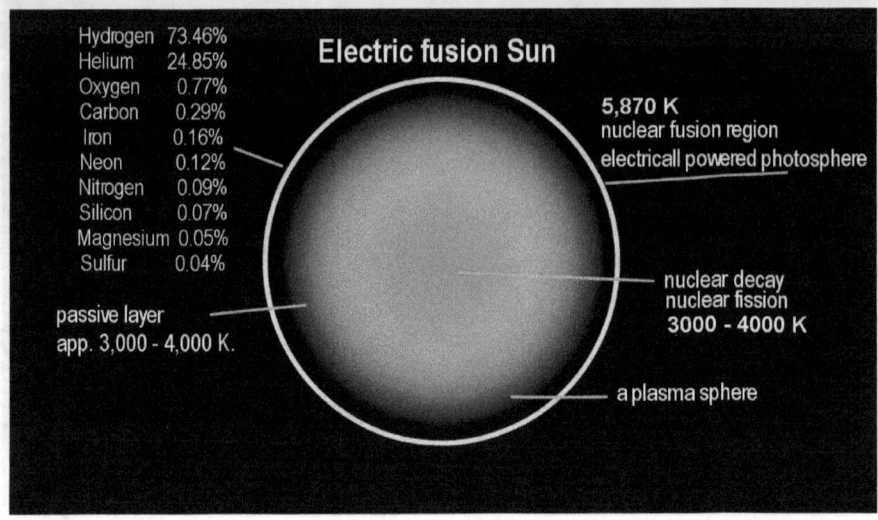

When plasma is motivated by the electric force, which is 39 orders of magnitude stronger than mass and gravity, chains of highly efficient nuclear fusion reactions occur. They occur right on the surface of the Sun, where the plasma meets the Sun's reaction cells. In the fusion reactions on the surface of the Sun, all the atomic elements are synthesized that the planets are made of. The synthesized atoms then flow away with the solar wind. During the early phase of the Sun, all the atoms for the planets were created and carried in the solar winds, till they fell out and condensed into planets.

This in short, is how the solar system was created near the center of the galaxy, and how it still functions fundamentally, though with lesser intensity.

Operational differences between the Sun and the Earth

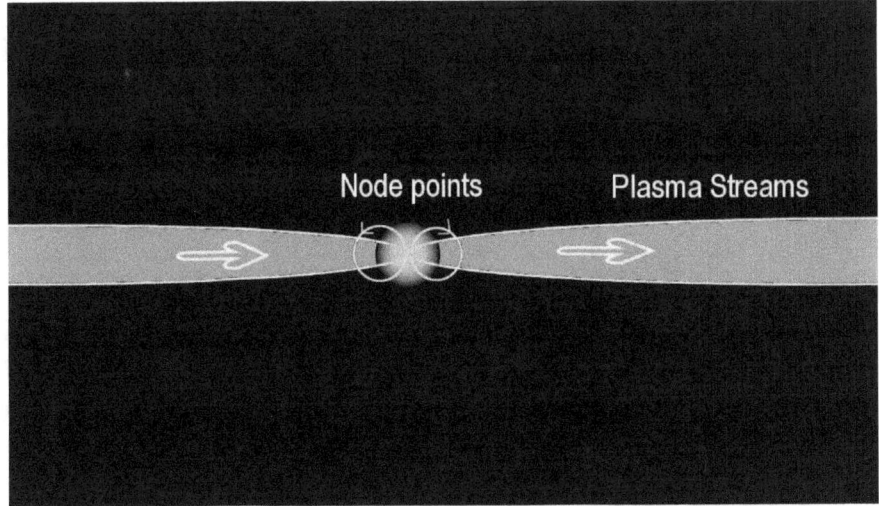

Of course, we also encounter some operational differences between the Sun and the Earth. For the intense plasma compression around the Sun to happen, which enables surface nuclear fusion to occur, large plasma streams are required to achieve the compression. They are needed to produce the necessary strong magnetic fields. Massive electric movement creates strong magnetic fields,

Plasma streams that connect stars

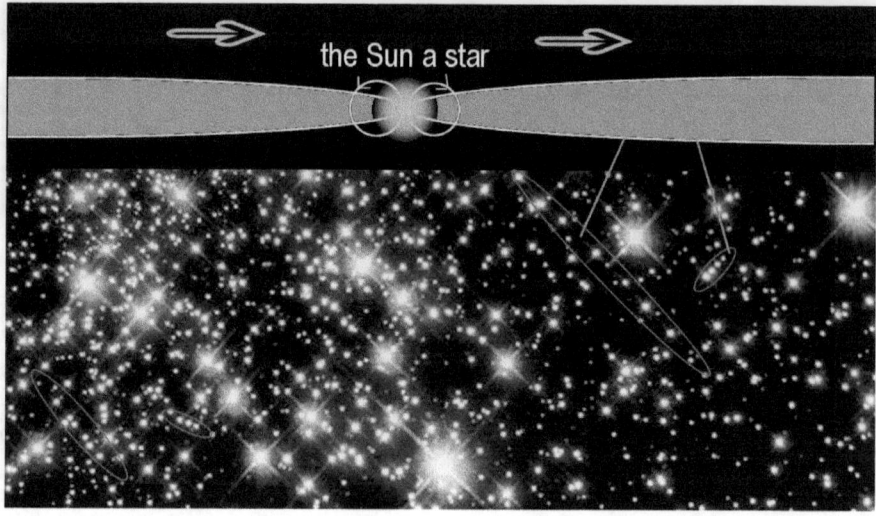

The interstellar plasma streams that connect stars in the galaxy, typically carry a greater volume of plasma than a Sun can consume with its nuclear-fusion synthesis. The excess plasma then simply flows on and away from the Sun in an outgoing plasma stream. While the streams flowing towards the Sun become magnetically pinched into ever-tighter confinement, the weaker streams flowing away from the Sun, begin tightly, and then expand. They typically pick up plasma along the way. They continue to expand till the density becomes large enough again for them to contract by magnetic pinching as they flow towards the next star, at the next node point.

The evidence for this dynamic interconnection between the stars is found in the typical alignment of stars into short and long strings. This string-like alignment is also visible in the fields of galaxies.

An internally powered sun is not needed

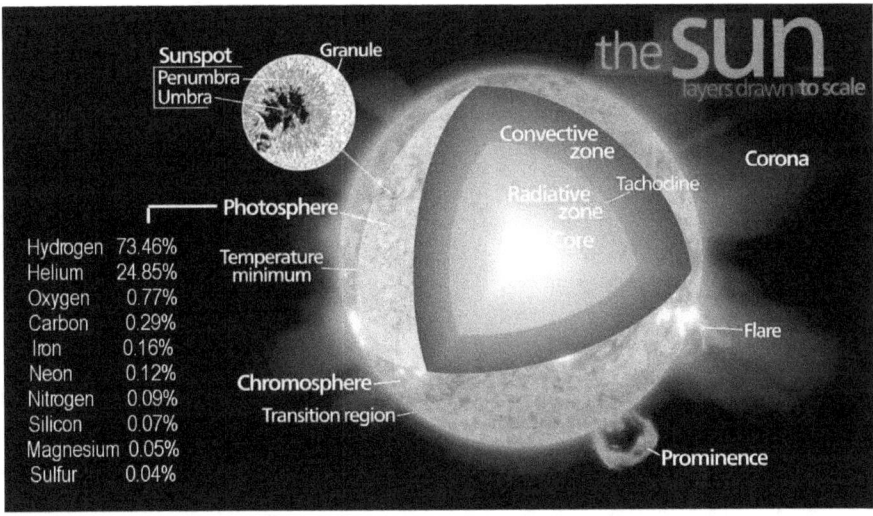

All this means that an internally powered sun is not needed. In fact, the internally powered Sun cannot perform the services for the universe that its dynamics require. The theory of the internally powered Sun simply doesn't work. It is so full of holes that it can't possibly work. Of course, it will likely take some time before the defective model of the Sun will be let go in the world of science and in public perception. Still, this needs to happen. The breakout to reality is urgently needed.

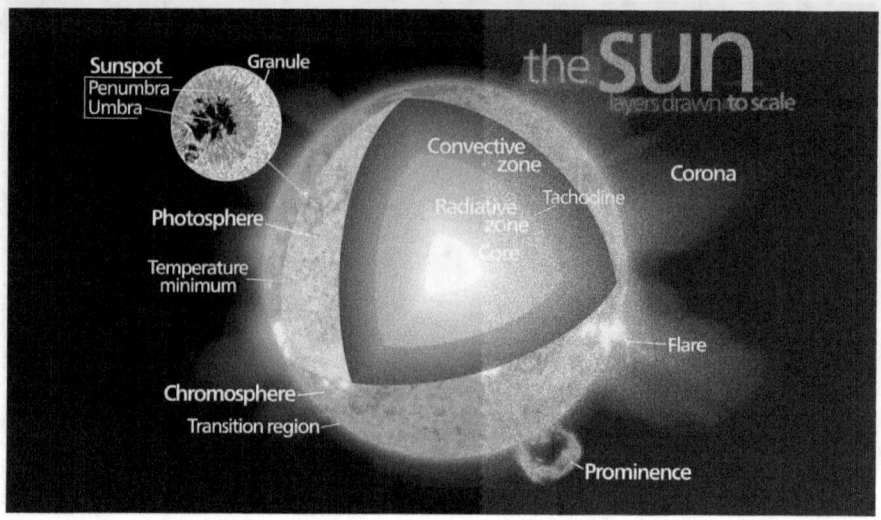

So what is wrong with the old theory about our Sun? Why is the Sun not a sphere of hydrogen gas? What is the generally accepted theory saying to us, that is wrong?

The general perception is that the Sun is a sphere of atomic gas, mostly hydrogen and some helium, and that within its core, at a calculated temperature of 15 million degrees Kelvin, at a gas density 150 times the density of water, nuclear fusion reactions do occur that generate vast amounts of energy by which the Sun becomes a brilliantly radiant sphere in the sky that has burnt for billions of years with unchanging intensity, and will keep on burning for a few more billion years to come.

The Sun is theorized to fuse hydrogen atoms together

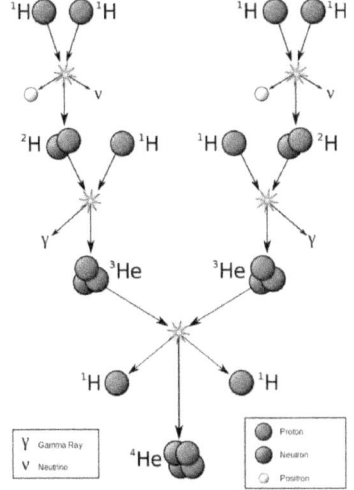

The theorized p-p (propon - proton) fusion chain

that is deemed to power the internal-fusion Sun

The Sun is theorized to fuse hydrogen atoms together in a chain of reaction by which helium atoms are produced and some energy. The produced energy-flux by mass is deemed to be equal to the energy generated by the metabolism of a human being. The extreme energy of the Sun is deemed to be the result of energy accumulation derived from its great volume.

That's the prevailing theory.

The Sun is anything but a constant star

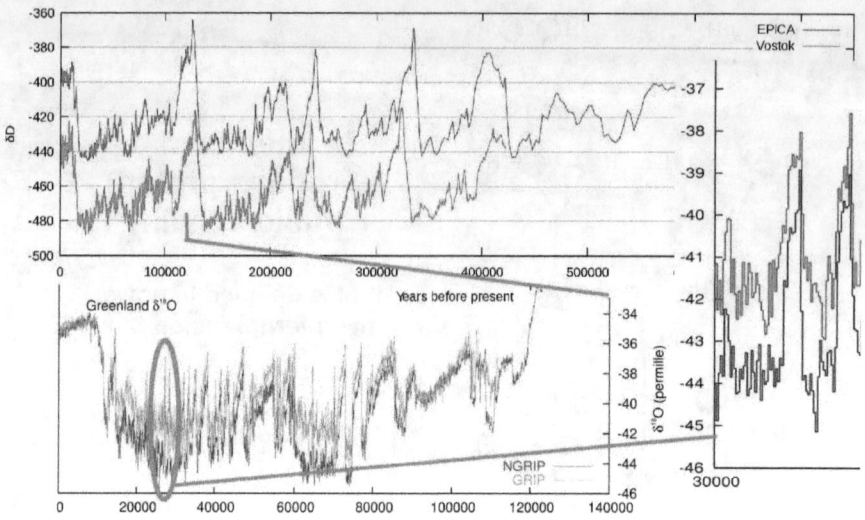

But the theory doesn't add up, does it? The long-term climate history that has been discovered, the big ice age cycles that we have evidence of, and even the ice core records of the last Ice Age with its gigantic short-term temperature oscillations that are evidenced in them, all tell us that the Sun is anything but a constant star. The internal solar fusion theory does not allow for the enormous short and long-term fluctuations that we have evidence of.

The heat generated at the core of the Sun

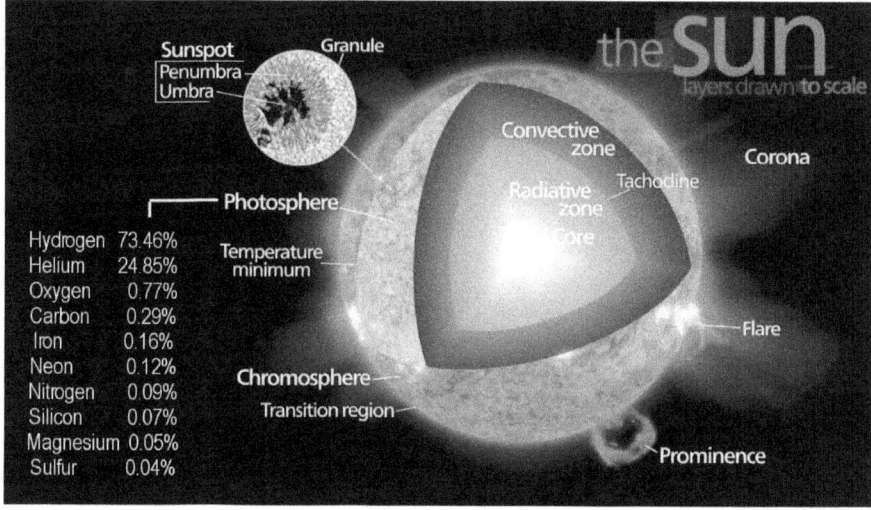

The theory tells us that the heat generated at the core of the Sun, by nuclear fusion, takes a path of 30 million years to slowly ooze to the surface by convection, and that even the heat transmission by photons takes 10,000 to 170,000 years to reach the solar surface. By this slow process, the Sun should be rock-solid, unvarying. But it isn't.

The Sun's energy cycle is oscillating

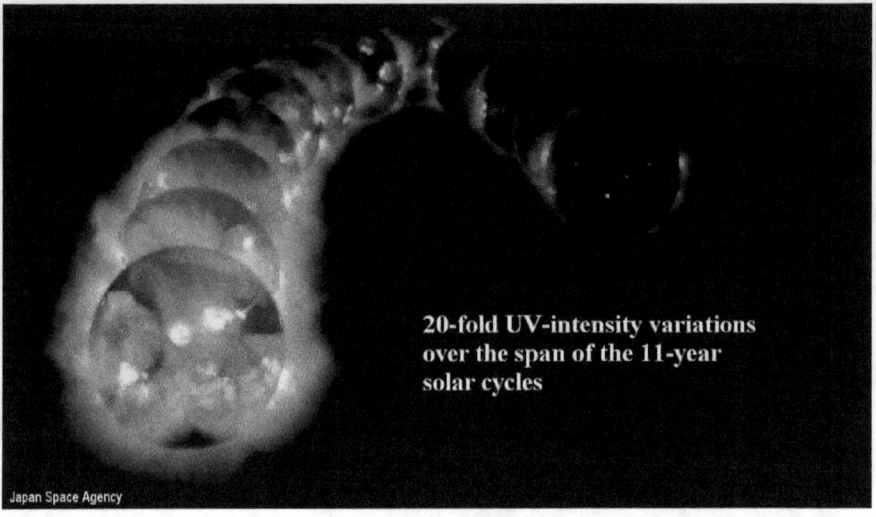

20-fold UV-intensity variations over the span of the 11-year solar cycles

Japan Space Agency

Even now, the Sun's energy cycle is oscillating at an 11-year beat. Although its visible light output remains presently steady within seven-tenth of a percent, in other parts of the spectrum its energy radiation, varies by a factor of 20, that's a two thousand percent difference.

There are so many items of physical evidence coming to the surface that shouldn't happen under the internal-fusion theory, but which do happen, that one is forced to conclude that the widely accepted theory is evidently wrong.

It shouldn't be possible for the solar wind

http://www.zam.fme.vutbr.cz/~druck/Eclipse/ — an example of the amazing solar eclipse photography of Miloslav Druckmueller

For example, it shouldn't be possible for the solar wind to accelerate against the force of gravity, as the wind flows away from the Sun.

The corona around the Sun is hotter than the Sun

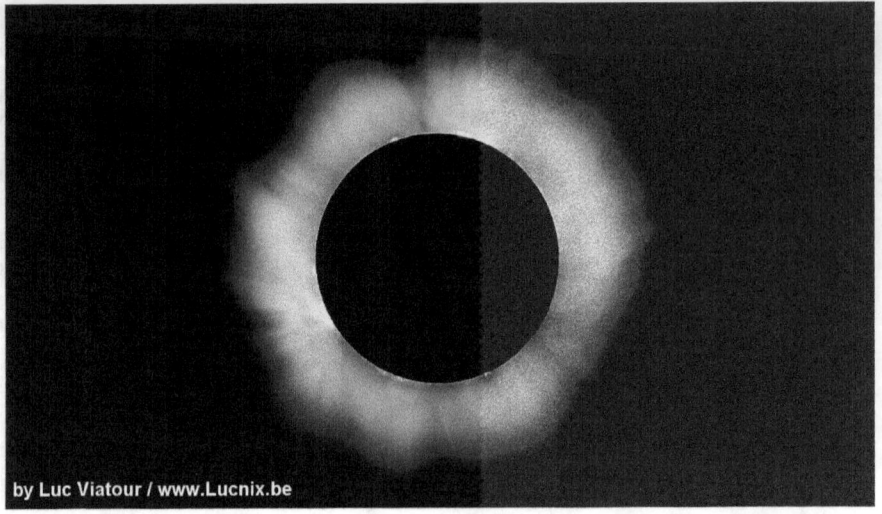

by Luc Viatour / www.Lucnix.be

Likewise, it shouldn't be possible that the corona around the Sun is hundreds of times hotter than the Sun itself. However, these impossibilities all happen. The theoretical impossibilities of self-evident facts, create paradoxes.

When we look at the sunspots on the Sun

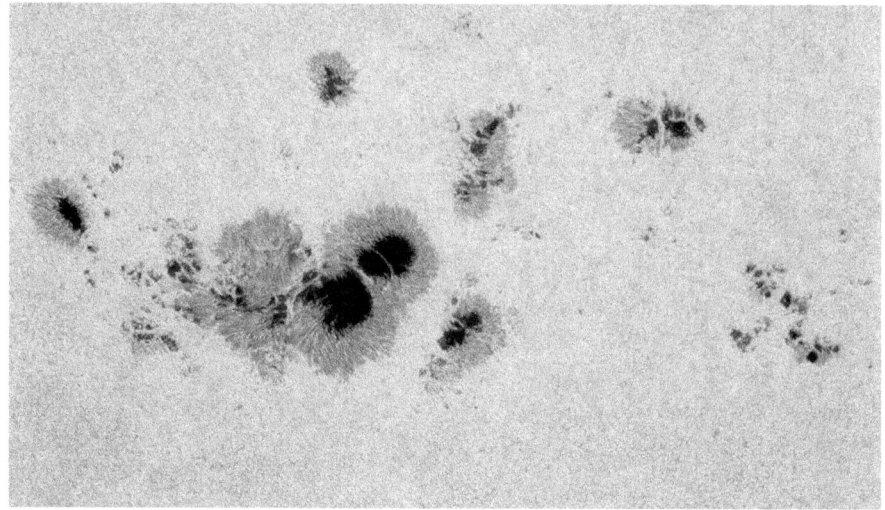

Neither should it be possible that when we look at the sunspots on the Sun, and look through the umbra below the surface, that the Sun is dark inside instead of being brilliantly radiant. However, as you can see for yourself, below its shiny skin, the Sun is dark.
Is this a paradox? No, it isn't. What we see is precisely what the Sun should be like, because nothing else is possible.

Impossible for the Sun to be a gas sphere

It is impossible for the Sun to be a gas sphere,
which the internal nuclear-fusion theory requires.

It is impossible for the Sun to be a gas sphere,
which the internal nuclear-fusion theory requires.

The mass-density of the Sun

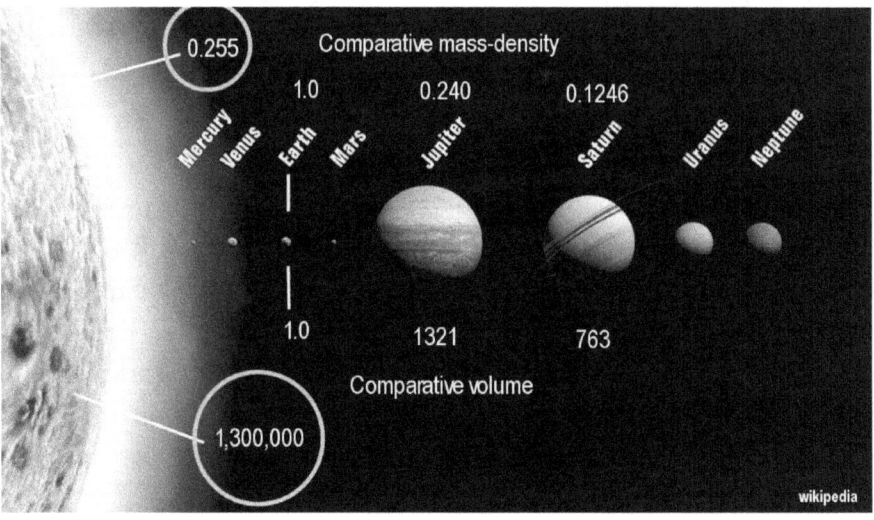

By what is required for the standard solar theory to function, the Sun should not exist. It is simply not possible for a gas-sphere of the size of the Sun, to exist. The gravity pressure in a gas sphere of such an immense size would be so great that all atoms in its core would be crushed, instead of atoms being fused into larger atoms. A gas sphere of the size of the Sun would also be a thousand times heavier than the Sun is known to be. No rational theory, no matter how exotic, can bridge this impossible paradox.

Just do some simple comparison. The gas planet Jupiter, at double the volume of Saturn, has double the mass density of Saturn. The doubling of the mass-density results from the greater compression of the gas by the greater gravity, which the larger gas-volume generates. This is what one would expect.

The Sun, in the same comparison, has a thousand times the volume of Jupiter, but it is known to have roughly the same mass-density as Jupiter. That's not possible.

The mass-density of the Sun would be more than a thousand times greater if it was a gas sphere. However, if the Sun was a plasma sphere, which is diffused by the repelling electric force that is inherent in plasma, the Sun's 'measured' mass-density is just about right.

The resulting gigantic paradox renders the internal fusion Sun theory, which depends on atomic hydrogen being fused into helium, to be totally wrong.

Great efforts have been made to explain the paradox away with exotic excuses, in order to rescue the false theory for which no evidence exists. Unfortunately, the process of shrouding paradoxes with exotic epicycles, for which no evidence exists either, is like saying to society, "we really don't know how the thing works. We are guessing. The paradoxical theory, impossible as it is, is the best we can come up with."

*The premise that 99.999% of the universe does not exist

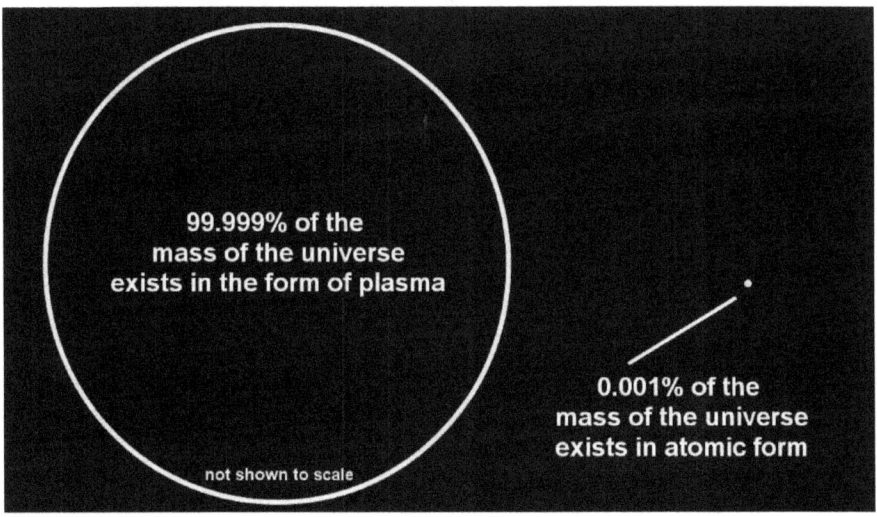

Of course, the paradoxical theory is the best theory possible when one is forced to proceed from the premise that 99.999% of the universe does not exist, so that the theory of the functioning of the Sun must be built on the one-thousandth of a percent of the universe that is allowed to be acknowledged in general perception. The result is necessarily, a hopelessly narrow perception, the kind that is characteristic of fairy tales that are told to children at bed time to put them asleep.

Plasma is the blood of the universe

Researchers at the Los Alamos National Laboratory have come to the conclusion that the Universe is not as empty as the fairy-tale scripts make it out to be.

The researchers have come to the recognition that 99.999% of the mass of the universe - which is the portion that the fairy tales do not include - does indeed exist, and exists in the form of plasma.

Plasma is the blood of the universe. It consists of the basic particles that all atoms in the universe are made of. The particles also exist in free-flowing form with a ratio way above 10,000 to 1.

While free-flowing plasma in space is invisible, as its particles are 100,000 times smaller than the smallest atom, the existence of giant plasma streams in cosmic space is discernable by their effects.

The effects of the plasma streams in space

For example, in cosmic space, we see the effects of the plasma streams in space in the alignment of stars and galaxies, often lined up in neat rows like beads on a string.

Secondary effects of flowing plasma

In the solar system we see the secondary effects of flowing plasma in the form of planets orbiting in an ecliptic around the Sun.

Plasma streams as solar wind

© Miloslav Druckmuller/Barcroft

http://www.zam.fme.vutbr.cz/~druck/Eclipse/ — an example of the amazing solar eclipse photography of Miloslav Druckmueller

We also see the effect of plasma streams as solar wind, which is in part why plasma is called the lifeblood of the universe. Plasma is the lifeblood of the Sun. It enables electric nuclear fusion at its 'surface' layer. The solar wind is the result of it.

The Sun is a plasma star
with electric nuclear fusion occurring on its surface that operates at low temperature

The Sun is a plasma star
with electric nuclear fusion occurring on its surface that operates at low temperature. No other types of solar energy are physically possible, nor would there be a need for other types, since electric plasma fusion on the surface of the Sun is the most efficient type of nuclear fusion possible. It is so efficient that it operates at such low temperature as the Sun's current 5,505 degrees Celsius.

The principle is efficient

The principle is efficient because its process is driven by electric interaction - by the interaction of one of the strongest forces in the universe.

Two main types of plasma particles

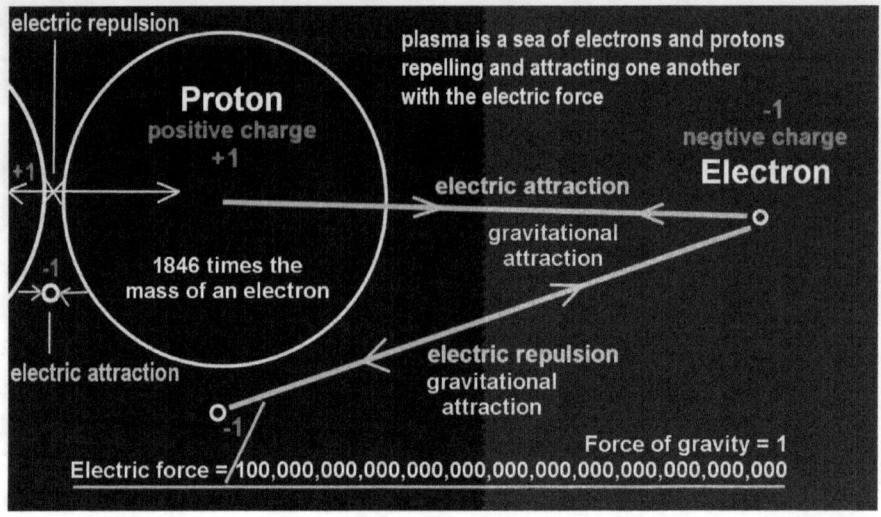

electric repulsion

plasma is a sea of electrons and protons repelling and attracting one another with the electric force

Proton
positive charge
+1

1846 times the mass of an electron

-1
negtive charge
Electron

electric attraction

gravitational attraction

electric repulsion
gravitational attraction

electric attraction

+1

-1

-1

Force of gravity = 1
Electric force = 1/100,000,000,000,000,000,000,000,000,000,000,000,000,000

Plasma exists in the form of extremely small energized particles that are deemed to be themselves but constructs, of constructs of energy. They carry both a quantity of mass, and a specific electric charge.

There are two main types of plasma particles recognized, a large type and a small type. The small type is named an electron. It carries a negative electric charge. The large type is named a proton. It is a thousand times larger than the electron and has nearly 2,000 times more mass, and it carries a positive electric charge.

As particles of mass, plasma particles gravitate towards one another by the effect of their mass, called gravity. The particles also affect one-another by the electric force. Particles that are of complimentary polarity, meaning opposite polarity, such as protons and electrons, attract one another by the electric force that is 39 orders of magnitude stronger than the force of gravity. Unlike, by the effect of gravity, plasma particles do also repel one-another by the electric force. This happens between particles of like polarity.

Electrons rebound

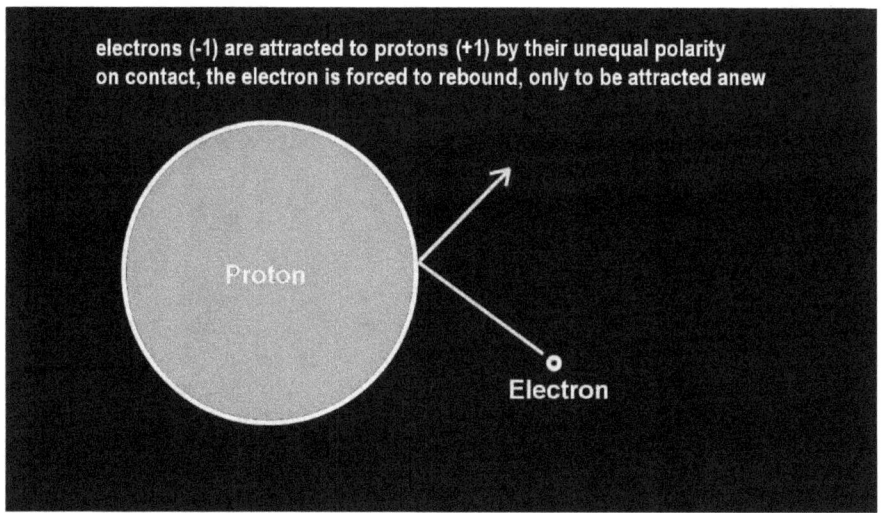

electrons (-1) are attracted to protons (+1) by their unequal polarity
on contact, the electron is forced to rebound, only to be attracted anew

Proton

Electron

However, the opposite happens when the attracting particles come extremely close to one another. The reason is, that the universe would not function without this additional principle.

For example, if electrons were merely attracted to protons, they would latch onto one-another. As a result, nothing would move. The electrons would simply remain stuck there. Nothing useful would result. But this doesn't happen. Nowhere in the universe can we find electrons latched onto protons. Before they would latch on, the electrons rebound.

Instead of getting stuck by being latched onto protons, at close distances, the electrons bounce away from the protons, like a rubber ball rebounds when it hits a wall. By this rebounding effect, electrons become drawn into an endless dance around the protons in plasma. They are attracted from afar, and repelled at close distances, only to become attracted anew.

A density determined zone

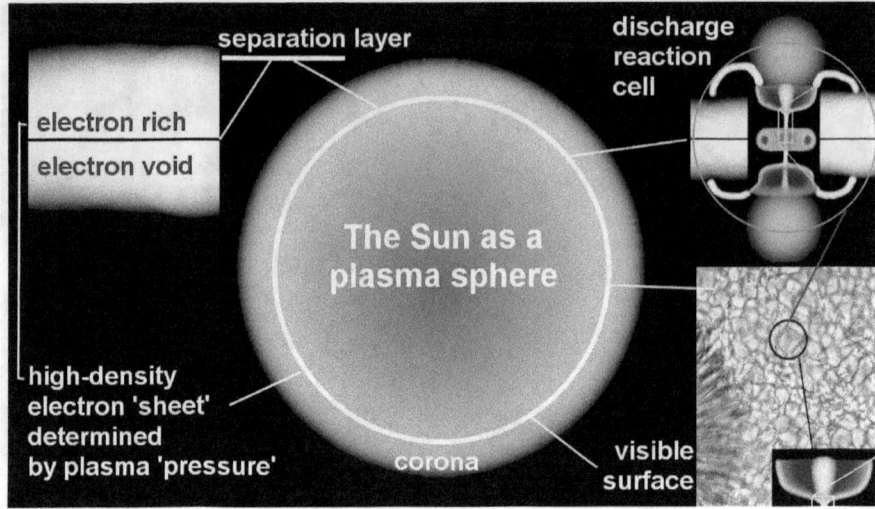

The same principle, may on the large scale, cause a repulsive barrier to form against the electron-dance, past a specific plasma density, which repels electrons out of the dense plasma region, somewhat like oil floating on water. This means that the visible surface of the Sun that we see, isn't actually a surface in the standard sense, but is merely a density determined zone where all the plasma interactions take place, including the fusion reactions that require a high electron-density.

A creative atom synthesizing 'engine'

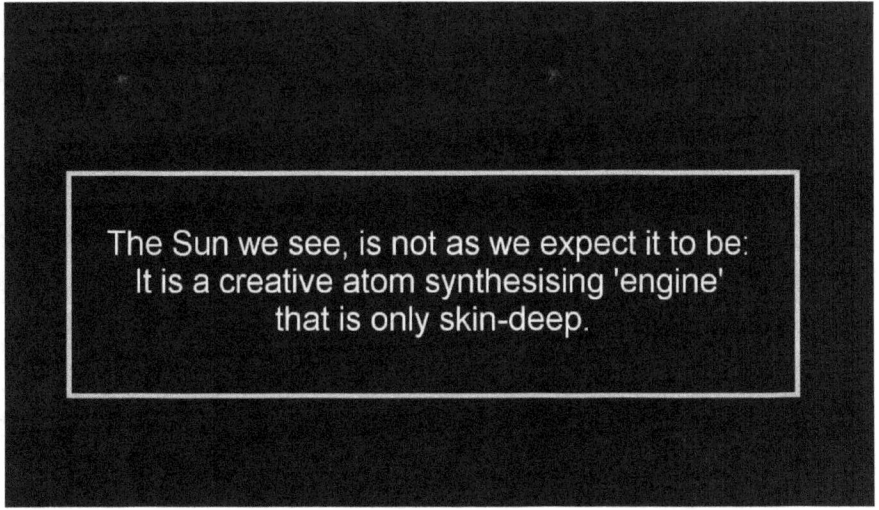

The Sun that we see, is not as we expect it to be: It is a creative atom synthesizing 'engine' that is only skin-deep.

When we look at the Sun

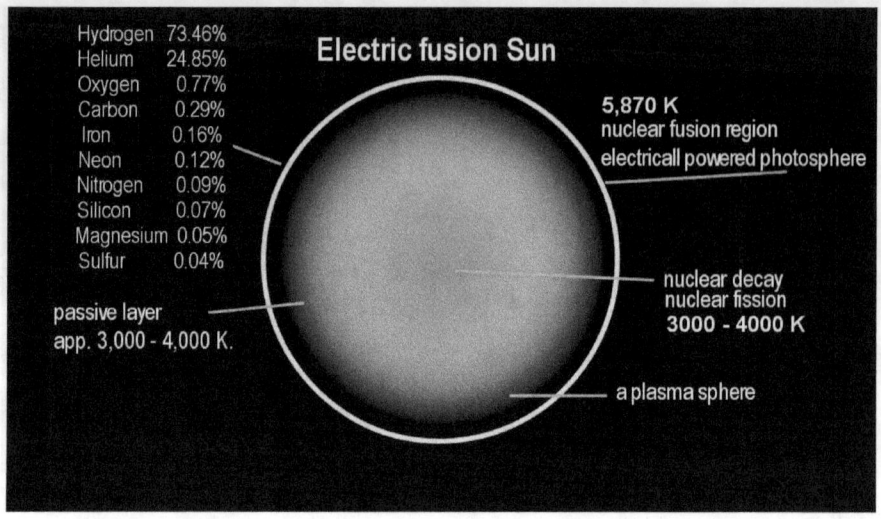

This means that when we look at the Sun, what we see, is in reality quite different than we imagine it to be. What we see is a thin active layer where electric interaction occurs, in which also nuclear fusion takes place, with an inactive ball of plasma below the active layer.

At the core of the plasma ball, some nuclear decay processes may occur that result when atomic structures drift down from the surface and become crushed by the plasma pressure, which then revert back to the plasma state from which they were created on the surface.

When the inflowing plasma stream becomes weak

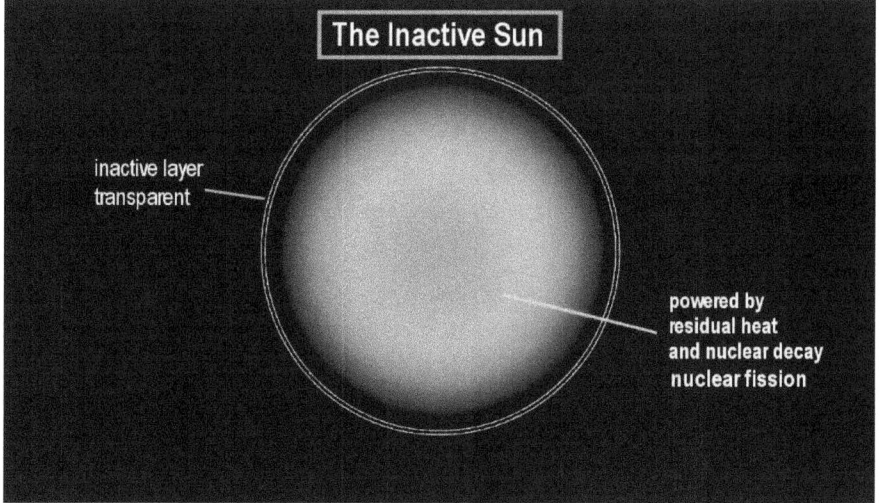

When the inflowing plasma stream becomes weak, and becomes insufficient to maintain the required electron density for the electric nuclear fusion to occur, the reactive layer will cease to function. It will simply cease to exist. A faint glow may remain as some weak reactions may continue to occur. And at the deep center a bright zone may be found, caused by the fission reactions of nuclear decay. In its inactive state, our Sun will likely become a 'white dwarf.'

Nuclear fusion happens on the surface of the Sun

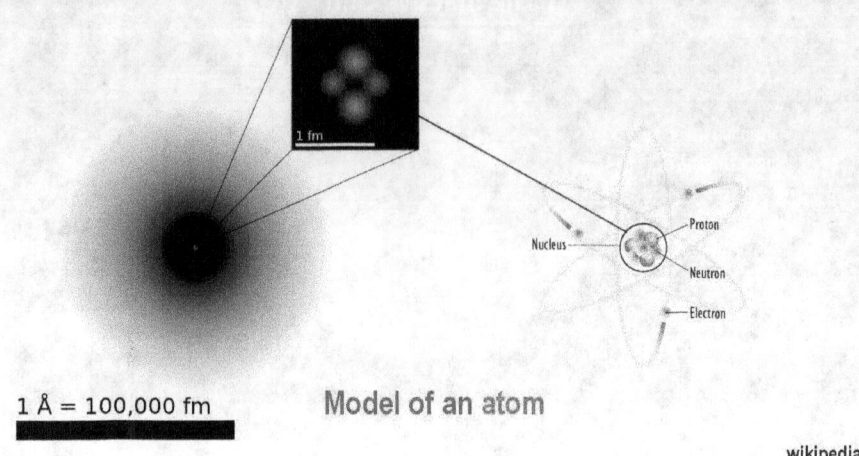

1 Å = 100,000 fm   Model of an atom

Nuclear fusion happens on the surface of the Sun, for as long as the Sun operates in a healthy, electron rich, environment. When electrons become highly energized, the rapid movements of their dance around the proton has the effect that the electrons seem to be everywhere. When this dance happens with a great energy, an atom is being born. A hydrogen atom results from a single electron 'swarming' a single proton. The atom that is formed by this highly energetic dance, is typically 100,000 times larger than the proton at the center of the dance, and is a million times larger than the swarming electron itself that gives the atom its effective form.

The dynamic dance of electrons

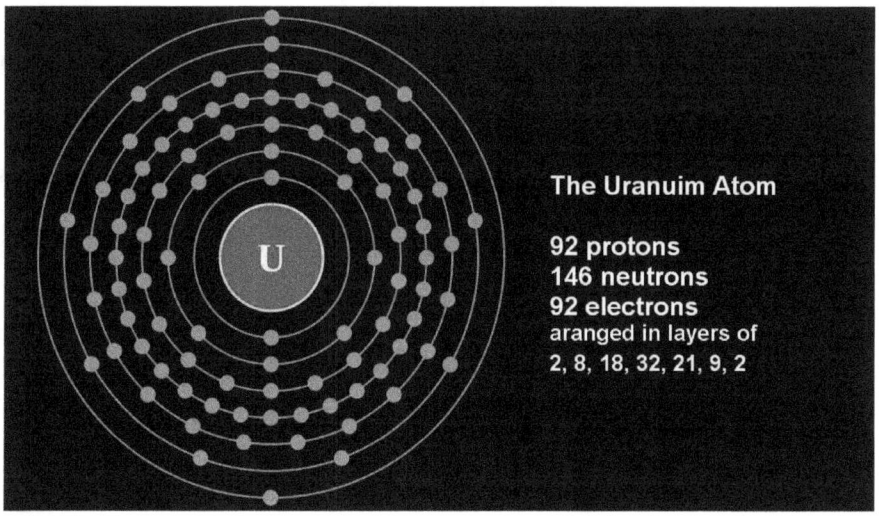

The Uranuim Atom

92 protons
146 neutrons
92 electrons
aranged in layers of
2, 8, 18, 32, 21, 9, 2

The dynamic dance of electrons, can be a swarm of one, or of more that a hundred swarms of electrons arranged into 'rooms' and layers of rooms, up to 7 levels deep, all contained within a single atom, which, altogether give the atom its form.

It may be the powerful electromagnetic effect in flowing plasma, accelerated by magnetic fields, which accelerates the protons to the velocity needed for them to fuse. The fusion cells on the surface of the Sun are typically up to 1000 kilometers wide. They operate as very large particle accelerators that fuse plasma into such large structures as the uranium nucleus that required the fusion of 238 protons, some of which become converted into neutrons in the process of fusing.

By the intense electric interactions of plasma in the fusion process and with its synthesized atoms, electromagnetic energy, both in the form of light and thermal energy, is being emitted.

When two protons are forced closely to each other

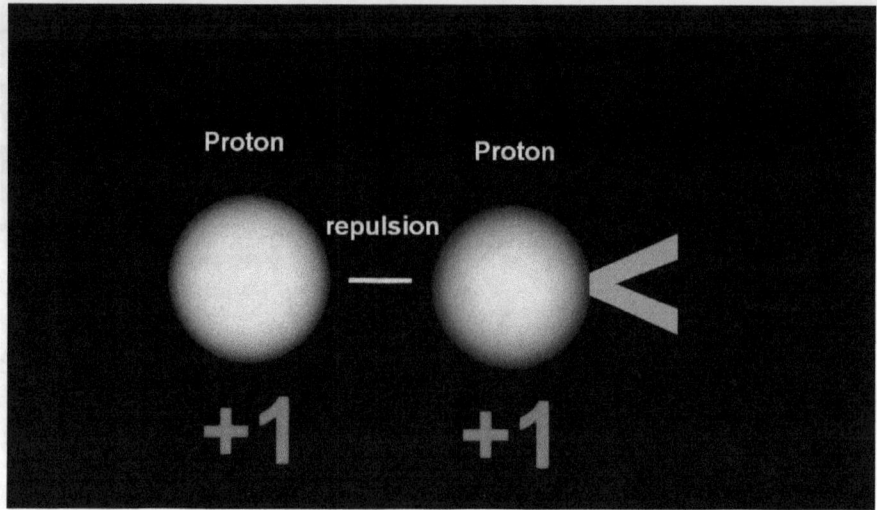

A different type of close-encounter effect occurs when two protons are forced closely to each other, with a greater force than their repelling electric force. In this case, at an extremely close distance, the protons' electric repelling force is reversed, and becomes an attractive force instead, which snaps the two protons together.

The neutron, essentially acts like a glue

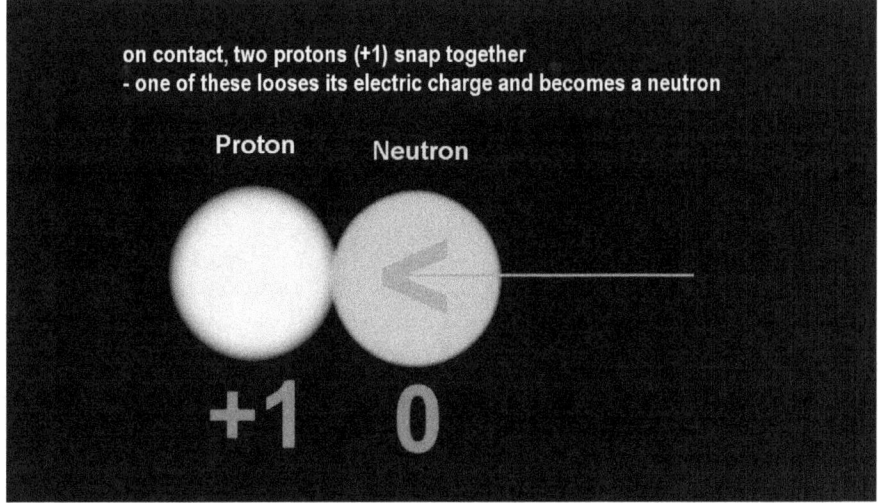

on contact, two protons (+1) snap together
- one of these looses its electric charge and becomes a neutron

Proton   Neutron

+1   0

The energy that overcomes the repulsion for this to happen, of course, has to be absorbed. It becomes absorbed by one of the protons, that thereby looses its electric polarity. The joined proton becomes electrically neutral in the process. It becomes a neutron. By this radical transformation of the previous proton into a neutron, the fused particles form a larger unit in which the neutron, essentially acts like a glue between protons in the resulting nucleus that becomes the center of an atom.

By the process of protons becoming bundled

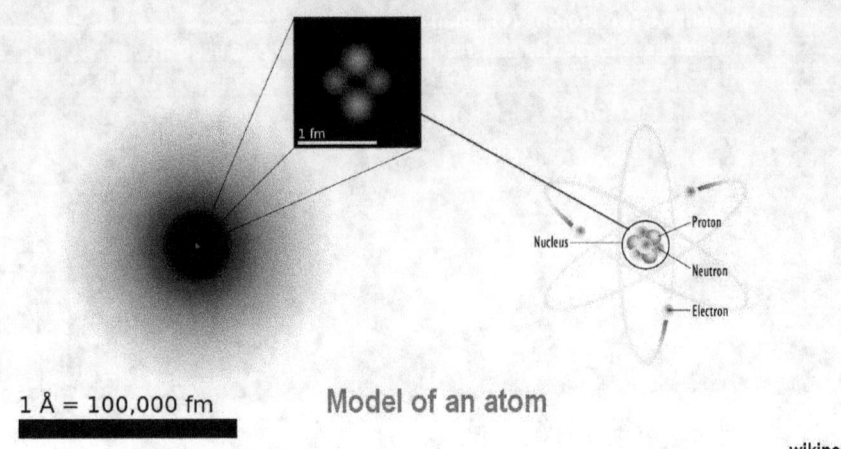

1 Å = 100,000 fm   Model of an atom

By the process of protons becoming bundled together, it becomes possible for large atoms to be formed that have numerous protons at their center, forming large clusters for large atomic structures, as in the case of uranium, where the entire large nucleus is being latched together with more neutrons that serve as glue for the giant package, than protons in the package.

A hundred different elementary atoms

Group →	1	2	3	4	5	6	7	8	9	10	11	12	13	14	15	16	17	18
↓ Period																		
1	1 H																	2 He
2	3 Li	4 Be											5 B	6 C	7 N	8 O	9 F	10 Ne
3	11 Na	12 Mg											13 Al	14 Si	15 P	16 S	17 Cl	18 Ar
4	19 K	20 Ca	21 Sc	22 Ti	23 V	24 Cr	25 Mn	26 Fe	27 Co	28 Ni	29 Cu	30 Zn	31 Ga	32 Ge	33 As	34 Se	35 Br	36 Kr
5	37 Rb	38 Sr	39 Y	40 Zr	41 Nb	42 Mo	43 Tc	44 Ru	45 Rh	46 Pd	47 Ag	48 Cd	49 In	50 Sn	51 Sb	52 Te	53 I	54 Xe
6	55 Cs	56 Ba		72 Hf	73 Ta	74 W	75 Re	76 Os	77 Ir	78 Pt	79 Au	80 Hg	81 Tl	82 Pb	83 Bi	84 Po	85 At	86 Rn
7	87 Fr	88 Ra		104 Rf	105 Db	106 Sg	107 Bh	108 Hs	109 Mt	110 Ds	111 Rg	112 Cn	113 Uut	114 Fl	115 Uup	116 Lv	117 Uus	118 Uuo

Lanthanides	57 La	58 Ce	59 Pr	60 Nd	61 Pm	62 Sm	63 Eu	64 Gd	65 Tb	66 Dy	67 Ho	68 Er	69 Tm	70 Yb	71 Lu
Actinides	89 Ac	90 Th	91 Pa	92 U	93 Np	94 Pu	95 Am	96 Cm	97 Bk	98 Cf	99 Es	100 Fm	101 Md	102 No	103 Lr

In this manner, more than a hundred different elementary atoms are being created on the solar surface. We find them arranged in the periodic table of elements, according to the number of protons contained in each package.

Typically, the ratio of protons to neutrons is 1 to 1. In larger atoms, a greater ratio of neutrons is required to hold the big nuclei together. As I said, more glue is needed there. However, in every case, the number of electrons in an atom, matches the number of protons, whereby the electric fields inside an atom balance each other out. By this perfect equality, every resulting atom becomes electrically neutral. While it is possible for the swarming electrons of closely spaced atoms to share each other's space, which latch atoms together into molecules and so on, the electric neutrality of the atomic structures remains always intact.

The balancing act that creates electrically neutral packages, which all atoms and molecules are, is of critical importance for the universe to exist, and for us to exist. This electric balancing act is the 'heart' that motivates everything. The electric balancing that creates electrically neutral atoms is one of the main factors why the cold-fusion Sun theory is possible, or for that matter, any atomic synthesizing fusion is possible.

Even the hydrogen atom is, heavy

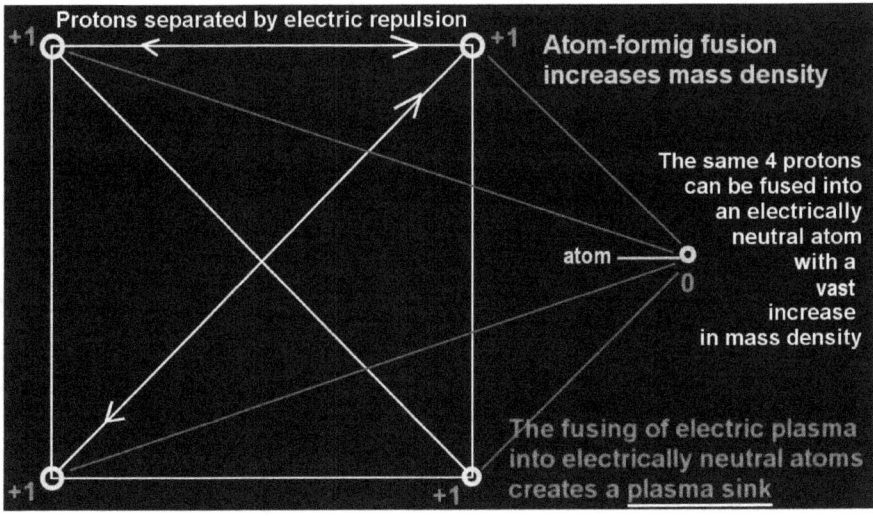

When an atom is formed that is electrically neutral, the repelling force that would keep the protons repelled from each other in plasma, is suddenly neutralized. This means that far-more protons, when joined to form atoms, can be packed into a given space than would otherwise be able to exist in that space.

In the unbound state, electric repulsion makes plasma extremely light and thinly diffused. In comparison, plasma bound into atoms is heavy, because the resulting package is dramatically smaller. Even the hydrogen atom is, heavy, in comparison with plasma, and of course, the helium atom is four-times heavier still.

The dense packaging of protons into atoms renders the internal solar nuclear-fusion theory, a fundamental impossibility.

Because of the tight packing of atoms, a gas sphere the size of the Sun, filled with atoms of hydrogen and helium, would likely be a thousand times heavier, if not more so, than the Sun actually is.

The Sun synthesises all the atoms in the solar system. What created the hydrogen atoms it is deemed to be made of?

As if the mass-density paradox was not enough to disprove the theory of the Sun being a hydrogen star, one may further consider that an even greater paradox exists. Where would the hydrogen have been produced that under the internal fusion theory, makes up a Sun? That's a paradox.

It is a paradox, because a sun is the only operating platform in the universe that causes the synthesis of atoms from plasma, including the hydrogen atom. The paradox is that the hydrogen could not have come from the Sun. A sun cannot produce itself. It cannot operate before it exists. Nor would the supposed accretion of dust separate out the light hydrogen for the Sun, and the heavy atoms for the planets. Accretion doesn't work that way.

The resulting built-in hydrogen origin paradox is so great that it closes the door on any possibility for the Sun to be a gas star with internal, atomic fusion happening in its core. This renders the theory as but a dream - a dream enabled by another dream, the Big Bang creation dream.

www.ingramcontent.com/pod-product-compliance
Lightning Source LLC
Chambersburg PA
CBHW060404190526
45169CB00002B/752